Mathematica Demystified

Demystified Series

Mathematica Demystified

Jim Hoste

New York   Chicago   San Francisco   Lisbon   London
Madrid   Mexico City   Milan   New Delhi   San Juan
Seoul   Singapore   Sydney   Toronto

The McGraw·Hill Companies

Library of Congress Cataloging-in-Publication Data

Hoste, Jim.
    Mathematica demystified / Jim Hoste.
        p.   cm. – (Demystified)
    Includes bibliographical references and index.
    ISBN 978-0-07-159144-7 (alk. paper)
    1. Mathematica (Computer file) 2. Mathematics–Data processing.   I. Title.
QA76.95.H67   2008
510.285′536–dc22   2008037478

Mathematica Demystified

1   2   3   4   5   6   7   8   9   0   DOC/DOC   0   1   3   2   1   0   9   8

ISBN   978-0-07-159144-7
MHID   0-07-159144-3

 The pages within this book were printed on acid-free paper containing 15% post-consumer fiber.

Sponsoring Editor
    Judy Bass
Acquisitions Coordinator
    Alexis Richard
Editing Supervisor
    David E. Fogarty
Project Manager
    Somya Rustagi, International
    Typesetting and Composition
Copy Editor
    Surendra Nath, International
    Typesetting and Composition

Proofreader
    Nigel Peter O'Brien, International
    Typesetting and Composition
Indexer
    Broccoli Information Management
Production Supervisor
    Richard C. Ruzycka
Composition
    International Typesetting and Composition
Art Director, Cover
    Jeff Weeks

To Mary, Benjamin, and Graeme

ABOUT THE AUTHOR

Jim Hoste, Ph.D., is a mathematics professor at Pitzer College and has used *Mathematica* since its initial release in 1988. He has been an associate editor of the *Journal of Knot Theory and Its Ramifications* since 1991 and has authored dozens of research publications.

CONTENTS

Contents

Contents

PREFACE

- **What Is *Mathematica*?**
  Mathematica is a program for doing mathematics. Using *Mathematica* it is easy to

  - Make numeric and symbolic calculations
  - Simplify complicated mathematical expressions
  - Plot the graphs of functions as well as curves and surfaces in 3-space
  - Create sophisticated color graphics
  - Compute derivatives and integrals
  - Solve equations, including differential equations
  - Work with large data sets
  - Create animations
  - Write programs to carry out any algorithm
  - Create slideshow presentations

  The latest version of *Mathematica*, Version 7.0, is packed with all sorts of new capabilities including a dynamic interface that allows the user to interact with graphics, as well as other kinds of output in real time. The new version of *Mathematica* is also fast. Many of the underlying routines have been optimized for greater speed. Whether you are a high school student or a Ph.D. mathematician, physicist, or engineer, *Mathematica* is an ideal tool for meeting your computational needs.

- **Who Is This Book for?**
  This book is intended primarily for *Mathematica* newcomers—people who have never used *Mathematica*, or who have used it very little. We'll take you from your very first calculation all the way to plotting beautiful fractals. The book includes hundreds of examples each with step-by-step explanations. Using this book, you can progress from knowing nothing at all about

Mathematica to being able to use *Mathematica* for all of the tasks listed above and more. Experienced *Mathematica* users who have yet to learn all the new features of Versions 6.0 and 7.0 should also find the book helpful.

While we assume you know nothing about *Mathematica*, we do assume that you know a little mathematics, at least high school algebra. A high school student who has yet to learn calculus may find the chapter on calculus a bit mysterious, but there are plenty of neat things to learn in the other chapters. College students who are currently taking calculus, linear algebra, or differential equations can use this book as a companion text, and do your homework with *Mathematica*.

In addition to teaching you *Mathematica*, the examples will introduce you to some beautiful and exciting mathematics. The book provides glimpses of some real mathematical gems including a few conjectures that have vexed mathematicians for hundreds of years. So, you'll not only learn *Mathematica*, but some neat mathematics too!

- **What's in This Book?**

Mathematica is *huge*, really, really *big*. Any book can cover only a small fraction of *Mathematica*. Our goal is to cover enough material so that after reading this book you can do quite a few things with *Mathematica*, probably as much as most people would ever want to do. Plus, you don't have to read the whole book. If, for example, you mainly want to use *Mathematica* to produce graphics, you can skip the later chapters that don't deal with graphics. But perhaps more important than teaching you a specific set of *Mathematica* features, we'll be teaching you how to use the online *Mathematica* documentation so that you can teach yourself to go beyond the basics of this book. Each chapter contains a section titled "Find Out More" that will direct you to guides and tutorials in the *Mathematica* Documentation Center that elaborate on the material in that chapter.

For the most part, each chapter deals with a specific topic within *Mathematica*, for example, three-dimensional graphics, or solving equations. The one exception is Chapter 10 that instead focuses on a mathematical topic, in this case, dynamical systems and fractals, and shows how to use *Mathematica* to draw beautiful fractals including Julia sets and the famous Mandlebrot set.

With only one exception, the chapters are ordered so that each *Mathematica* function is thoroughly explained when it is first introduced and before it is used in later chapters. Still, it is not necessary to read the chapters in the given order. However, if you are a complete beginner, you should at least read Chapters 1 to 5 (or perhaps at least 1 to 4) before skipping around. You could probably read the chapter on three-dimensional graphics right after the chapter on two-dimensional graphics, especially if you are willing to flip back to earlier sections in the book occasionally to fill in some missing details. The last chapter will show you how to add text and organize your *Mathematica Notebooks* so that they are more than mere scratch sheets for your calculations. If you are a student who plans to do mathematics homework with *Mathematica*, don't put off reading this chapter!

Each chapter closes with a Quiz, and a Final Exam can be found at the end of the book. The Quiz and Exam questions vary from fairly easy to quite challenging. Try to do them yourself before looking at the answers!

Jim Hoste
Claremont, CA
July, 2008

ACKNOWLEDGMENTS

A number of people have offered helpful comments and suggestions. Thanks are due to Judith Grabiner for several of the historical remarks in the book. Robert Fierro offered useful feedback when the project was just beginning. Paul Wellin from Wolfram Research shared some sample notebooks that were inspirational. Judy Bass, at McGraw Hill, provided valuable guidance and support.

CHAPTER 1

Getting Started

1.1 Starting *Mathematica*

When we start *Mathematica* a fresh window, or *notebook*, will open. This is where we will do all of our mathematical calculations and graphics. At the top of the window we see the title of the window, which initially is "untitled-1." Later, when we see how to save our work as a file, we can give a name to the file, and that name will appear in the title bar of the window.

1.2 Entering Expressions

Let's do our first calculation! If we type 1+1 and then press Shift+Return (i.e., hold down the Shift key and then the Return key)[1] *Mathematica* computes the sum and places the answer on the next line in the window. This is called *evaluating* or *entering* the expression. The window now contains

---

[1]On Mac keyboards, Shift+Return is the same as Enter. With either Windows or Mac OS, using the Return key will simply move the cursor to the next line, allowing us to type more.

Example 1.2.1

In[1]:= **1 + 1**

Out[1]= 2

Notice that *Mathematica* has placed "In[1]:=" and "Out[1]=" labels to the left of 1+1 and 2, respectively. To the right of the input and output, *Mathematica* has placed a set of *brackets*. The two innermost brackets enclose the input and output, respectively, and the larger bracket groups the input and output together. Each bracket contains what is known as a *cell*. All of the calculations that we do in this notebook will be organized into cells and the brackets that surround the cells will come in handy for organizing our work. We'll have a lot more to say about this in Chap. 11, so don't worry too much about the brackets now. In fact, *until we get to Chap. 11 we will be omitting the brackets most of the time when we display Mathematica input and output.*

1.3 Editing Cells

Let's change 1+1 to 1+2. *Mathematica* supports all the usual mouse-driven text-editing features of word processors. We can simply use the mouse to place the cursor in the input cell and edit the entry so that it reads 1+2. To redo the calculation, we now reenter the cell by once again pressing Shift+Return. The result is

---

Example 1.3.1

In[2]:= **1 + 2**

Out[2]= 3

Notice that the the In and Out labels have changed to "In[2]:=" and "Out[2]=." Each time we reevaluate a cell, the numbers in the In and Out labels will change.

To create a new cell with a new calculation, simply start typing. *Mathematica* will place the input in a new cell. When many cells are present we can use the mouse to place the cursor between existing cells and click the mouse button to insert a new cell at that location. Notice how the cursor changes from a vertical bar when located inside a cell, to a horizontal bar when located between cells. With the cursor between cells, click the mouse button and then start typing. *Mathematica* will create a new cell at the desired location.

Finally, we can also click on the bracket which encloses a cell to *select* it. After selecting a cell we can reevaluate it by pressing Shift+Return or treat it just like any selected item in a text document and cut, copy, or paste as usual. Try deleting an entire cell by clicking on its bracket and then choosing **Edit ▶ Cut** from the menu bar, (or using the equivalent keyboard shortcut).

There are lots of ways that we can change the appearance of cells, changing the font, fontsize, color, and the like. We'll explore these topics in Chap. 11.

1.4 Basic Arithmetic

Mathematica can do all the basic operations of addition, subtraction, multiplication, division, and exponentiation (raising one number to another) which are denoted by the symbols $+$, $-$, $*$, $/$, and $\wedge$.[2] We can also use parenthesis for grouping as usual. Here is an example involving the arithmetic operations.

Example 1.4.1

In[3]:= **2 ∗ 3 + 4 ∧ 2**

Out[3]= 22

Here the exponentiation was done first, giving $2*3+16$, then the multiplication, which leads to $6+16$, and finally the addition. *Mathematica* follows the standard order of operations, first performing all exponentiation (from left to right), then all multiplications and divisions (again from left to right), and finally, all additions and subtractions (from left to right). If we want to override these conventions we need to use parenthesis to group terms.

One nice feature of *Mathematica* is that of *implied multiplication*. We do not need to use the multiplication sign $*$ in order to multiply. Instead, a blank space between things that can be multiplied (numbers, variables, expressions) will be treated as multiplication. The blank space can even be omitted if parenthesis are used to indicate multiplication. If we do leave a blank space for multiplication, sometimes *Mathematica* will fill in the space with the multiplication symbol $\times$. Basically, we can type calculations pretty much the way we would write them. Here are several examples, all contained in a single input cell.

---

[2]When computers were first introduced, exponentiation was denoted by the "up-arrow" ↑. The shaft of the arrow was eventually lost and we were left with only the arrowhead.

Example 1.4.2

In[4]:= 5×6
$2 (3 + 4)$
$(2 - 3 + 1) (1 + 2 / 3) - 5\,\hat{}\,(-1)$
$6!$

Out[4]= 30

Out[5]= 14

Out[6]= $-\dfrac{1}{5}$

Out[7]= 720

Here we entered four separate calculations in a single input cell. (This is when you use the Return key—to type a new line in the input cell.) Notice that each result is placed in its own output cell. We didn't use the multiplication sign for $5*6$ in the first calculation and instead left a blank space. After entering the blank space and the 6, *Mathematica* inserted the $\times$. In the second and third calculation, because of the parenthesis, there is no confusion caused by leaving out the multiplication sign, so it is easier not to use it. The fourth calculation illustrates the *factorial* symbol !. We read 6! as "six factorial" rather than shouting SIX. By definition, $n!$ is the product of all integers from 1 to n. Thus $6! = 6 \cdot 5 \cdot 4 \cdot 3 \cdot 2 \cdot 1 = 720$.

1.5 Using Previous Results

Quite often we will perform a calculation and then want to use the output of this calculation for our next calculation. We can use the percent symbol, %, to refer to the output of the previous cell. Here is an example.

Example 1.5.1

In[11]:= $2\,\hat{}\,5$

Out[11]= 32

In[12]:= $\% + 100$

Out[12]= 132

Notice that the first cell gave output of 32 and that the next cell added 100 to this to give 132. In this case the % symbol referred to the previous output. We can even use %% to refer to the result before the last result, or even %%% for the result before that. Sometimes using the % symbol can be quite handy. However, it is important to remember that % *always refers to the last output*. This can sometimes lead to unexpected results! In this book, we will rarely use the % symbol.

1.6 Exact versus Approximate

One of the truly amazing features of *Mathematica* is that it will work things out *exactly* whenever possible. Sometimes this is just what we need, but sometimes it would be nicer to get an approximate answer. Consider the following example.

Example 1.6.1

In[10]:= **3 ^ 20 / 2 ^ 21**

Out[10]= $\dfrac{3\,486\,784\,401}{2\,097\,152}$

It's pretty hard to get a feel for the fraction $\frac{3486784401}{2097152}$ and it might be nicer to approximate it with a decimal representation. We can force *Mathematica* to do this in two important ways. The first is to use decimal representations from the very beginning. If we replace 3^{20} with 3.0^{20} (or $3^{20.0}$, or even $3.0^{20.0}$) look what happens.

Example 1.6.2

In[11]:= **3.0 ^ 20 / 2 ^ 21**

Out[11]= **1662.63**

Mathematica always views decimal representations as approximations. Thus *Mathematica* considers 3.0 to be an approximate number rather than an exact number. If we ever do a calculation that involves approximate numbers, *Mathematica* will give an approximate answer. On the other hand, if we use exact numbers in the input, *Mathematica* will do its best to provide exact numbers in the output. Here are several more examples that illustrate this point.

Example 1.6.3

In[12]:= **3 / 4**

    3.0 / 4.0

    12 ^ (1 / 2)

    12 ^ .5

Out[12]= $\dfrac{3}{4}$

Out[13]= 0.75

Out[14]= $2\sqrt{3}$

Out[15]= 3.4641

Notice that 12^(1/2) is the square root of 12 and that this is *exactly* equal to $2\sqrt{3}$. So *Mathematica* has not only given us an exact answer, it has also simplified the input. On the other hand, by replacing the exponent of 1/2 by the "approximation" of .5 we have forced *Mathematica* to give us an approximate answer in decimal form.

The second important way to force *Mathematica* to give approximate answers is to use the *numeric evaluation* function **N**. We describe this function in the next section.

1.7 Using Functions

Mathematica has thousands of built-in functions. Fortunately, we only have to know a few dozen[3] of the more important ones to do lots of neat calculations. We will be introducing the most important and useful functions in this book as we go. The next example uses the square root function **Sqrt** and the numeric evaluation function **N**.

Example 1.7.1

In[16]:= **Sqrt [27]**

    N [Sqrt [27]]

Out[16]= $3\sqrt{3}$

Out[17]= 5.19615

---

[3]OK, I lied. Knowing a hundred functions would be nice. Actually, memorizing the names of most functions is not so hard. How hard can it be to remember **Cos** for cosine, **Abs** for absolute value, and **Total** for, well, total? The real work is going to be remembering the *syntax* needed to use these functions. A good strategy will be to get good at using the built-in documentation.

Here we see the use of the built-in square root function **Sqrt**. Entering Sqrt[27] is the same as entering 27^(1/2). In the first computation we entered the number 27 exactly, so obtained an exact answer. In the second calculation we still entered 27 exactly but *Mathematica* provided an approximate answer. We forced this to happen by using the numerical evaluation function **N**. This function will convert any number into a decimal representation. Many functions in *Mathematica* have *optional arguments* and the numerical evaluation function **N** is one of them. By adding the optional argument n, **N**[x, n] will estimate x with n-digit precision. If we want to know a certain number of digits in the decimal representation of π, for example, we can use **N** as in the first line of Example 1.7.2.[4] Here we use **Pi** to stand for π, the ratio of the circumference to the diameter of any circle. *Mathematica* has special symbols for a number of important mathematical constants, including **E**, the base of the natural logarithm, and **I**, the imaginary number whose square is -1.

Example 1.7.2

```
In[33]:= N[Pi, 100]
      N[Pi / 10, 30]
      N[Pi / 10.0, 30]

Out[33]= 3.141592653589793238462643383279502884197169 ⋅
      399375105820974944592307816406286208998628 ⋅
      34825342117068

Out[34]= 0.314159265358979323846264338328

Out[35]= 0.314159
```

The second and third lines of Example 1.7.2 illustrate an important feature of **N**. In both cases we have asked *Mathematica* for 30 digits in the expansion of $\pi/10$. In the first case we get it, but in the second case we do not. This is because in the second case we have already moved to an approximation by using 10.0 instead of 10.

Here are some sample calculations involving the constants **E** and **I**.

---

[4]You may remember that π is an *irrational* number, one whose decimal representation never ends and never repeats. Amazingly, in November of 2005, Chao Lu recited the first 67890 digits in the decimal expansion from memory! Check out http://www.pi-world-ranking-list.com/index.html.

Example 1.7.3

In[21]:= `Sqrt [−16]`
     `N [E, 10]`
     `E ^ (I Pi)`

Out[21]= 4 i

Out[22]= 2.718281828

Out[23]= − 1

The last calculation is one of the more amazing identities in all of mathematics! It follows from Euler's formula[5]

$$e^{i\theta} = \cos\theta + i\sin\theta.$$

If we substitute $\theta = \pi$, we obtain $e^{i\pi} = -1$.

In the last computation we tried to take the square root of -16 and *Mathematica* responded with the imaginary number $4i$. *Mathematica* is perfectly happy using complex numbers $a + bi$ where a and b are real numbers and i is the imaginary number $\sqrt{-1}$. The numbers a and b are called the *real* and *imaginary* parts of $a + bi$, respectively. *Mathematica* has several built-in functions that deal especially with complex numbers. Two of the more important ones are the functions **Re** and **Im** which return the real and imaginary parts, respectively, of a complex number. Another important function is the absolute value function **Abs** which works not only for real numbers but for complex numbers too. In the case of a complex number $a + bi$, its absolute value is defined as $\sqrt{a^2 + b^2}$. Example 10.3.10 gives a few more calculations.

Example 1.7.4

In[24]:= `(2 + 4 I) (6 − 3 I)`
     `Re [2 + 4 I]`
     `Im [6 − 3 I]`
     `Abs [− 23]`
     `Abs [3 + 4 I]`

Out[24]= 24 + 18 i

Out[25]= 2

---

[5]Leonhard Euler (1707–1783) was one of the greatest mathematicians of all time. He published his famous formula in 1748.

Example 1.7.4 (Continued)

In[26]= − 3

Out[27]= 23

Out[28]= 5

Mathematica has all the common mathematical functions built-in. These include the trigonometric functions and their inverses, the hyperbolic trigonometric functions and their inverses, and the logarithm and exponential function. *Mathematica* also has many special, more esoteric functions too. In this book we will be primarily interested in the more common mathematical functions.

There are two *very important* features of all built-in *Mathematica* functions. First, *all built-in functions in Mathematica begin with capital letters*. Some, like the inverse cosine, **ArcCos**, may even have multiple capital letters. Second, *square brackets are always used to surround the input, or arguments, of a function.* So we type `Abs[-12]`, not `Abs(-12)`, if we want to compute the absolute value of −12. Moreover, this is the *only* use of square brackets in *Mathematica*. (Actually, the only use of *single* square brackets. We'll see shortly that *double* square brackets, [[and]], are used with lists.) Parentheses, (and), are used to group terms in algebraic expressions. One other set of *delimiters* that will be extremely important are the "curly braces," { and }. These are used to delimit lists, something that we will be introducing shortly. The three sets of delimiters, [ ], ( ), and { } are used for functions, algebraic expressions, and lists, respectively, and *only* for these purposes. This can be hard to get used to at first, but leads to a great system.

1.8 Using Variables

We may introduce variables and give them values using the equals sign. Here are some examples.

Example 1.8.1

In[29]:= **a = 2**

        b = 3

        a + b

Out[29]= 2

Out[30]= 3

Out[31]= 5

Now that we have defined a to be equal to 2, it will remain equal to 2 unless we set it equal to something else, or use the **Clear** function to clear its value. This is extremely important and can sometimes lead to a great deal of frustration! If we forget that we have given a value to a certain variable and then try to use the variable later as if it had no value, we can run into unexpected results. Example 1.8.2 shows how the **Clear** function works. Remember that previous to evaluating the cell, $a = 2$ and $b = 3$.

Example 1.8.2

In[32]:= `Clear[a]`

`a + b`

`Clear[b]`

`a + b`

Out[33]= `3 + a`

Out[35]= `a + b`

Variables that are given values retain those values until we quit *Mathematica* or use **Clear**. It is very important to remember this! Also, once we use **Clear** the variable will continue to *not* have a value until we give it one. Thus, if we reenter the above cell we will not get the same output! The second time we enter it, both a and b will have been cleared and we will not get the output of 3 + a. **Clear** can be used with a number of options. A useful construction is **Clear["Global`*"]** which will clear everything!

The real power of *Mathematica* is that it can manipulate abstract expressions rather than just specific numbers. So we will often use variables that are, well, variable! That is, they have not been set equal to any specific value. The following example illustrates this. We'll talk a lot more about the **Expand** function in Chap. 4, but for now you can probably guess what it does.

Example 1.8.3

In[36]:= `Expand[(x + y)^10]`

Out[36]= $x^{10} + 10\ x^9\ y + 45\ x^8\ y^2 + 120\ x^7\ y^3 +$
$210\ x^6\ y^4 + 252\ x^5\ y^5 + 210\ x^4\ y^6 +$
$120\ x^3\ y^7 + 45\ x^2\ y^8 + 10\ x\ y^9 + y^{10}$

Variables are *case sensitive*. Thus s and S are two different variables. Example 1.8.4 illustrates this point.

Example 1.8.4

In[29]:= a = 2

   b = 3

   c = 4

   s = a + b + c

   S

Out[29]= 2

Out[30]= 3

Out[31]= 4

Out[32]= 9

Out[33]= S

Since s (for sum) is the sum of a, b, and c, we see 9 for the fourth output line. But because S is not the same variable as s, the fifth output line contains the name of the variable S. This variable has no value since we have not set it equal to anything.

You can use almost anything as a variable name except that variable names cannot start with a number. Thus $x2$ can be used but $2x$ cannot. Moreover, words or letters that already have meaning in *Mathematica* cannot be used. For example, we cannot use **E** as a variable name because **E** is already being used by *Mathematica* to stand for the base of the natural logarithm. Other *reserved* words and letters exist too. If you try to use one, *Mathematica* will simply tell you that you are not allowed to. Here is an example of this.

Example 1.8.5

In[42]:= C = 12

       Set::wrsym : Symbol C is Protected. >>

Out[42]= 12

Since **C** is reserved we cannot use it for a variable, and *Mathematica* warns us that this is the case by typing the rather cryptic "Set::wrsym: Symbol C is

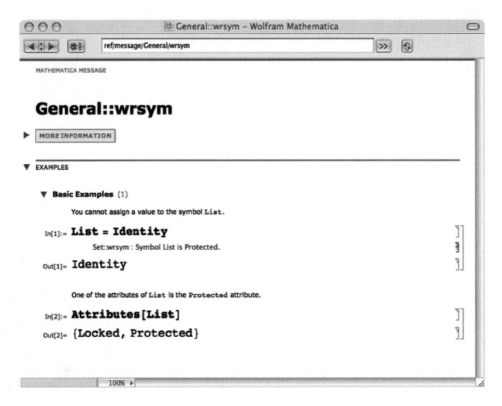

Figure 1.1 Trying to use the reserved word **C** as a variable causes a warning that leads to this page in the Help Files.

protected." Moreover, the double arrowhead, $>>$, is actually a hyperlink to the on-line documentation, or *Help Files*. If we click on this link the window shown in Fig. 1.1 pops up and explains the warning. We'll have a lot more to say about the Help Files as we go, starting a little later in this chapter.

1.9 Using Comments

After we start to do more complicated calculations, our input cells might start to have dozens of lines. When this happens, it can start to get hard to follow what is going on. Putting *comments* in our input cells, especially the more complicated input cells, is a great way to document our work. The following example illustrates the use of comments.

Example 1.9.1

```
In[43]:= (* distance from sun to earth in meters *)
    au = 149 597 870 691
    (* speed of light in meters per second *)
    c = 299 792 458
    (* time for light to reach earth from
     sun in seconds *)
    N[au / c]

Out[43]= 149 597 870 691

Out[44]= 299 792 458

Out[45]= 499.005
```

The delimiters (* and *) are used to enclose comments. Anything that appears between these delimiters is ignored by *Mathematica* when the cell is evaluated. Learning to use comments well is a very good programming practice and writing good code can be a source of great pride. Later we'll be talking about the *Wolfram Demonstrations Project*, a Web site that contains thousands of *Mathematica* notebooks that you can download for free. This is a great resource and someday you might find a notebook there that does almost exactly what you want to do. Excitedly, you'll download the notebook, open it up, see a hundred lines of mysterious code, and... What! No Comments! #@!&%* ! Good comments can make your code much better by making it readable by others (and by yourself after you have forgotten what you were thinking when you wrote it!).

In addition to comments, we can also add whole paragraphs of text between our input cells. We'll be talking about this in Chap. 11.

1.10 Suppressing Output

If we compute something that produces a LOT of output, we may want to hide or suppress the output just because it takes up so much room.

Example 1.10.1

In[39]:= `(* a very large Mersenne prime *)`

$x = 2^{4253} - 1$

`PrimeQ [x]`

Out[39]= 190 797 007 524 439 073 807 468 042 969 529 173 669 ⸜
356 994 749 940 177 394 741 882 673 528 979 787 005 ⸜
053 706 368 049 835 514 900 244 303 495 954 950 709 ⸜
725 762 186 311 224 148 828 811 920 216 904 542 206 ⸜
960 744 666 169 364 221 195 289 538 436 845 390 250 ⸜
168 663 932 838 805 192 055 137 154 390 912 666 527 ⸜
533 007 309 292 687 539 092 257 043 362 517 857 366 ⸜
624 699 975 402 375 462 954 490 293 259 233 303 137 ⸜
330 643 531 556 539 739 921 926 201 438 606 439 020 ⸜
075 174 723 029 056 838 272 505 051 571 967 594 608 ⸜
350 063 404 495 977 660 656 269 020 823 960 825 567 ⸜
012 344 189 908 927 956 646 011 998 057 988 548 630 ⸜
107 637 380 993 519 826 582 389 781 888 135 705 408 ⸜
653 045 219 655 801 758 081 251 164 080 554 609 057 ⸜
468 028 203 308 718 724 654 081 055 323 215 860 189 ⸜
611 391 296 030 471 108 443 146 745 671 967 766 308 ⸜
925 858 547 271 507 311 563 765 171 008 318 248 647 ⸜
110 097 614 890 313 562 856 541 784 154 881 743 146 ⸜
033 909 602 737 947 385 055 355 960 331 855 614 540 ⸜
900 081 456 378 659 068 370 317 267 696 980 001 187 ⸜
750 995 491 090 350 108 417 050 917 991 562 167 972 ⸜
281 070 161 305 972 518 044 872 048 331 306 383 715 ⸜
094 854 938 415 738 549 894 606 070 722 584 737 978 ⸜
176 686 422 134 354 526 989 443 028 353 644 037 187 ⸜
375 385 397 838 259 511 833 166 416 134 323 695 660 ⸜
367 676 897 722 287 918 773 420 968 982 326 089 026 ⸜
150 031 515 424 165 462 111 337 527 431 154 890 666 ⸜
327 374 921 446 276 833 564 519 776 797 633 875 503 ⸜
548 665 093 914 556 482 031 482 248 883 127 023 777 ⸜
039 667 707 976 559 857 333 357 013 727 342 079 099 ⸜
064 400 455 741 830 654 320 379 350 833 236 245 819 ⸜
348 824 064 783 585 692 924 881 021 978 332 974 949 ⸜
906 122 664 421 376 034 687 815 350 484 991

Out[40]= True

In Example 1.10.1, we compute $2^{4253} - 1$ which just so happens is a prime number![6] Here we have also used the function **PrimeQ** which will test an integer for primality.

There are two things we can do. First, we can simply hide the output by double-clicking on the cell bracket that surrounds both the input and output cells. If we double-click the bracket again it will redisplay the output. Try it! This way we can hide the output yet still have access to it if we need it. We won't be saying much more about cell brackets until we get to Chap. 11, but until then you should be comfortable with single-clicking a cell bracket to select that cell, and double-clicking brackets to hide or unhide large output cells.[7]

Alternatively, if we follow any calculation with a semicolon, the output from the calculation will not be displayed at all. For example, the semicolon following a=5^100 000 below will cause no output to be displayed, even though a will be given the value of 5^{100000}.

Example 1.10.2

```
In[38]:= (* semicolons suppress output *)
    a = 5 ^ 100 000;
```

Using semicolons also will allow us to place more than one command on the same line in the input cell. Here is a simple example.

Example 1.10.3

```
In[34]:= (* placing multiple commands on one line *)
    a = 2; b = 3;
    a + b

Out[35]= 5
```

We could have placed the sum a+b on the same line too, but using two lines makes for more readable input. Trying to develop good habits in style can be important as we learn how to do more and more complicated calculations.

Suppose we do a calculation that produces a lot of output and we don't remember to, or don't want to, suppress the output? Fortunately, *Mathematica* will step in and

---

[6]Primes of the form $2^n - 1$ are known as Mersenne primes after Marin Mersenne who compiled a (partially correct) list of them in the 17th century. So far, only 46 Mersenne primes have been found, with the largest having over 12 million digits!

[7]Double-clicking the bracket of any cell that is part of a larger group of cells will hide all the other cells in the group. Try it!

save us from having to look at pages and pages of output. For example, suppose we compute $12345^{1000000}$. Here is what happens.

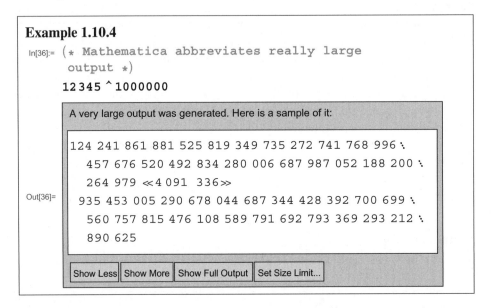

Example 1.10.4

In[36]:= (* Mathematica abbreviates really large
        output *)
        12345 ^ 1000000

A very large output was generated. Here is a sample of it:

124 241 861 881 525 819 349 735 272 741 768 996 ⟍
   457 676 520 492 834 280 006 687 987 052 188 200 ⟍
   264 979 ≪ 4 091 336 ≫
Out[36]= 935 453 005 290 678 044 687 344 428 392 700 699 ⟍
   560 757 815 476 108 589 791 692 793 369 293 212 ⟍
   890 625

| Show Less | Show More | Show Full Output | Set Size Limit... |

The beginning and end of the answer are displayed, with "<<4091336>>" appearing to indicate that the middle 4091336 digits of the answer are not being displayed. The number $12345^{1000000}$ has 4091492 digits and would take a lot of space to display! We are also given the choice to see more or less of the output.

1.11 Aborting a Calculation

Sometimes we might want to interrupt a calculation. For example, if we unknowingly start a calculation that might take days to finish, we'll just be waiting and waiting wondering how long it is going to take! Or, after we learn how to program in *Mathematica*, we might accidentally write a program that has a mistake in it that will cause the computer to run forever without ever completing what we wanted it to do. Rather than just quitting *Mathematica* and losing all of our work, we can usually abort the calculation by choosing **Evaluation ▶ Abort Evaluation** from the menu bar.

When *Mathematica* is doing a calculation it will say "Running ..." in the title bar of the window. If the calculation is really fast you won't even notice, but if the calculation lasts for several seconds you will see it. Try computing $5^{10000000}$. This calculation should last long enough for you to see the title of your window change to include "Running ...." This would be a great place to use a semicolon to suppress output! It is also a great place to use the **Timing** function so that we can see how long the calculation takes. The **Timing** function can be *wrapped* around any other calculation. Here is an example.

Example 1.11.1

In[51]:= (* Timing can be wrapped around any
       calculation *)
    Timing[5^10 000 000;]

Out[51]= {1.17752 , Null}

This calculation took 1.17752 seconds on the authors' laptop. The "Null" (which means "nothing") is present because we used the semicolon to suppress the output. Notice that the semicolon is *inside* the last bracket. We have simply taken the **Timing** function and wrapped it completely around the calculation that we want to do. Can you predict what will happen if we enter **Timing[5^100000000];**? Try it! In addition to using **Timing**, we'll see another way to monitor the speed of our calculations in Chap. 3.

1.12 Lists

Lists are so important to *Mathematica* that we need to learn what they are right away. In *Mathematica* a list is an ordered set of things delimited by { and } and separated by commas. Example 1.12.1 gives a few examples.

Example 1.12.1

In[17]:= (* examples of lists *)
    a = {1, 2, 3, 4, 5}
    days = {"Mon", "Tue", "Wed", "Thur",
       "Fri", "Sat", "Sun"}
    B = {{1, 2}, {3, 4}}

Out[17]= {1, 2, 3, 4, 5}

Out[18]= {Mon, Tue, Wed, Thur, Fri, Sat, Sun}

Out[19]= {{1, 2}, {3, 4}}

We have defined three lists and named each with a variable name. The name of the first list is **a**. This list has five *elements*, namely, the integers 1 through 5. The first element of the list is the number 1, the second element of the list is 2, and so on.

The second list contains the days of the week which are represented as *strings*. In general a string is any amount of text surrounded by quotation marks.[8] Finally, the last example is a list with two elements, each of which is a list itself! The elements of a list can be anything: numbers, variables, expressions, strings, or even other lists. The elements of a list do not have to all be the same "kind" of thing. It would be perfectly OK for a list to contain a number, a string, and perhaps another list.

Many of the functions in *Mathematica* use lists as one of their arguments. For example, the **Table** function, which is used to generate lists, takes two arguments, the second of which must be a list. Suppose we want to create a list of the squares of the first 10 integers. These are $1^2 = 1$, $2^2 = 4$, $3^2 = 9$, and so on up to $10^2 = 100$. Example 1.12.2 shows how to do this.

Example 1.12.2

In[55]:= (* a table of perfect squares *)
    Table[i^2, {i, 1, 10}]

Out[55]= {1, 4, 9, 16, 25, 36, 49, 64, 81, 100}

The **Table** function takes two arguments. In this case the first argument is the expression i^2. The second argument is the list {i, 1, 10}. What **Table** does is create a list by evaluating the expression i^2 with each value of i running from 1 to 10. The variable i is called the *index* or *counter*. It starts at 1 and is increased by 1 at a time until it reaches 10. For each value of i the expression is evaluated and the result is placed in the list. The argument, {i, 1, 10}, is extremely common in *Mathematica*. We will see arguments of this kind in many of the built-in functions provided by *Mathematica*.

Sometimes we may not want the counter to go up in steps of 1. We can change this by adding an optional *stepsize* to the indexing list. Example 1.12.3 illustrates different values for the stepsize. Can you see what is happening?

Example 1.12.3

In[56]:= (* examples of stepsize *)
    Table[i^2, {i, 1, 10, 2}]
    Table[j + 3, {j, 5, 1, −1}]

Out[56]= {1, 9, 25, 49, 81}

Out[57]= {8, 7, 6, 5, 4}

---

[8]Quotation marks from another set of delimiters.

In the first table, the counter i goes up in steps of 2, so that we only list the squares of the odd integers from 1 to 10. In the second table, the counter j goes *down* from 5 to 1 in steps of 1 (or *up* in steps of −1). The entries of the table, or list, are the index plus 3 so we obtain 8, 7, 6, 5, and 4.

In general, the indexing argument to the **Table** command is of the form

$$\{index, \ lower \ value, \ upper \ value, \ stepsize\}$$

although variations on this are possible. (For example, we may omit the stepsize in which case *Mathematica* will use the default stepsize of 1.) The index starts at the lower value, increases each time by an amount equal to stepsize, and ends when it reaches, or surpasses, the upper value. Notice that when we used $\{i,1,10,2\}$ the last value of i is 9. The next value would be 11 which is more than 10. So in this case the index is never actually equal to the upper value, instead it skips over the upper value.

We'll be learning a lot more about lists in Chap. 4, but there are a couple of list related functions that are worth mentioning now. Suppose we want to "access" a specific element of a list, say the third element. In the **days** list above, the third element is "Wed." The construction is to use the name of the list followed by double square brackets enclosing the number of the element we want. Thus, typing days[[3]] gives us the the third element of the list **days**. The list B in Example 1.12.1 was actually a list of lists. Hence, entering B[[1]] is the list $\{1,2\}$. If we want to access the second element of the first element of B we can type either B[[1]][[2]] or B[[1,2]]. We illustrate this in the following example.

Example 1.12.4

In[10]:= (* getting elements from a list *)
    days[[3]]
    B[[1]]
    B[[1, 2]]

Out[10]= Wed

Out[11]= {1, 2}

Out[12]= 2

Another useful function for dealing with lists is the length function **Length**. We can use this to find out how many elements are in a list. In Example 1.12.5 we use the **Table** command to first build a list and then the **Length** command to see how long it is. (Of course we *knew* in advance how long it was—this is just to illustrate using **Length**.) Also, notice that by following the **Table** command with a semicolon, *Mathematica* does not print out the list in an output cell.

Example 1.12.5

In[59]:= **cubes = Table[i^3, {i, 1, 10}];**
        **(* Length will give the number of
          elements in a list *)**
        Length[cubes]

Out[60]= 10

1.13 Palettes

So far we have seen how to enter the square root of 12 in two different ways: as **Sqrt[12]** or as **12^(1/2)**. A third way is to enter is as $\sqrt{12}$ by using the "Basic Mathematics Input Palette." If we select **Palettes ▸ BasicMathInput** from the menu bar, a window will open from which we may then select various forms of algebraic expressions, relational symbols, and Greek letters. Figure 1.2 shows what the Palette looks like. If we click on the square root expression, which is the first expression in the second row, the square root symbol, $\sqrt{\ }$, will be placed into the input cell and the cursor will be placed under the square root symbol so that we can begin typing there. After typing 12 we may press Shift+Return and obtain the answer. Example 1.13.1 shows how it looks.

Example 1.13.1

In[37]:= **(* using the square root symbol *)**

     $\sqrt{12}$

Out[37]= $2\sqrt{3}$

The BasicMathInput Palette contains a number of *templates* easily recognized by the little squares that are present, some of which are black. If you click on a Palette template it will be inserted in your notebook and whatever you type next will be inserted in the template at the location of the black square. Pressing the Tab key will take you to the next square in the template. If you select some expression and then click on a template in the Palette, whatever you selected will be pasted into the template at the location of the black square. The black squares are called *Selection Placeholders* while the white squares are simply *Placeholders*. After filling in all the place holders, type Ctrl+space (the Control key and the spacebar at the same

Figure 1.2   The Basic Math Input Palette.

time), or use the right arrow key, to move the cursor to the right of the template if you wish to keep typing.

If we want to enter $\sqrt{12} + \sqrt{27}$ we need to perform the following steps.

1. Select the square root symbol from the Palette.
2. Type 12.
3. Type Ctrl+spacebar.
4. Type +.
5. Select the square root symbol from the Palette again.
6. Type 27 and press Shift+Return.

Using the BasicMathInput Palette can make your *Mathematica* input cells look pretty, but this is primarily a typesetting feature. Almost anything that can be done by using the Palette can also be done without using it! Whether you end up using the Palette a lot or not is largely a matter of taste. On the other hand, really complicated expressions can be easier to "see" if they are typeset. So using the Palette to enter complicated expressions can be quite useful. A few of the templates available in the Palette can also be inserted by selecting **Insert ▶ Typesetting** from the menu bar. Furthermore, these have keyboard shortcuts that make using them a lot faster than selecting from the Palette. For example, to typeset $\sqrt{12} + \sqrt{27}$ using keyboard shortcuts we would type

Ctrl+2, 1, 2, Ctrl + spacebar[9], Ctrl+2, 2, 7

(Here the commas separate the keystrokes—don't type the commas!)

In addition to the keyboard shortcuts available on the **Insert** menu, the Greek letters all have shortcuts too. Typing esc, *letter*, esc (esc is the Escape key) will insert the Greek equivalent of *letter*. Thus esc, a, esc will insert α. Typing esc, e, esc will insert ϵ but typing esc, ee, esc will insert **e**, the base of the natural logarithm. Similarly, typing esc, ii, esc will give the imaginary number **i**. (The special constant π is typeset with esc, p, esc, not esc, pp, esc. You can't use π as a variable name. It *has* to represent the ratio of the circumference of a circle to its diameter.)

You'll need to experiment to see how much use you want to make of the Palette. Certainly the keyboard shortcuts make using it more palatable! We'll have more to say about using Palettes, including making your own custom Palettes, in Chap. 11.

---

[9] You may also be able to use the right arrow key.

1.14 Saving and Printing Our Work

To save our work, we choose **File ▶ Save** from the menu bar and give the name and location where the file should be saved. *Mathematica* files are known as *notebooks* and are stored with a ".nb" extension. Notice that after saving the file, the title appears in the title bar of the window where it used to say "untitled-1." It is a good habit to save our work often just in case the computer should crash for some mysterious reason. This may be a rare occurrence, but when it does happen we don't want to lose all our work!

It is possible to have many notebooks open at the same time and switch between them by using the **Window** menu. We'll have a lot more to say about notebooks in Chap. 11 where we'll see how to insert text, photos, and all sorts of other materials into a notebook and organize the whole work into chapters and sections just like a book.[10] Once we have created several notebooks, or downloaded notebooks from various Web sites, we may open them by selecting **File ▶ Open** from the menu bar. Most of the commands in the **File** or **Edit** menus will be familiar to anyone who has worked with other programs such as word processors. Still, some of the commands are not so obvious and will be covered in this book.

To print a notebook simply choose **File ▶ Print**. Notice that it is also possible to print a singe cell, or selection of cells. First click on the cell bracket to select a cell and then choose **File ▶ Print Selection**.

1.15 Getting Help!

Mathematica is equipped with a *huge* collection of files and tutorials that explain how to use the program. There must be thousands of pages, if not tens of thousands of pages, of documentation. These files can be accessed by choosing **Help ▶ Documentation Center** or **Help ▶ Virtual Book** from the menu bar. We will generally refer to this reference as the "Help Files" and it is very important for the *Mathematica* user to learn how to navigate and use the Help Files.

We have already seen one case of using them. Namely, when we tried to use the letter **C** as a variable and *Mathematica* gave us a warning. In that case, a hyperlink appeared in our notebook and if we clicked it we were taken to a page in the Help Files which explained the problem.

Alternatively, if we select **Help ▶ Documentation Center**, a window will open with lots of links for us to choose from. At the top of this window is a search field where we can type a word or phrase we want to find out about. If we type in a

---

[10]The people at Wolfram encouraged me to write this book as a *Mathematica* notebook!

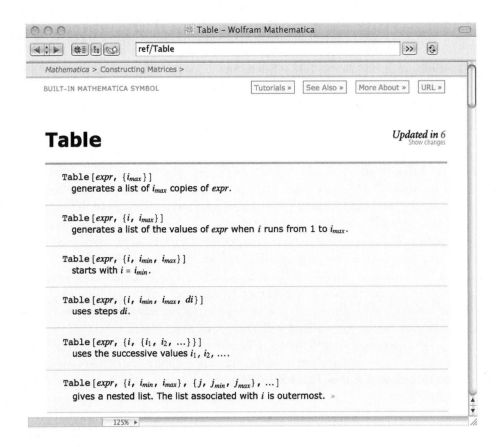

Figure 1.3 The help page on the Table command.

Mathematica command or function there we can get the documentation for that function. For example, suppose we type **Table** into the search field.

Figure 1.3 shows the Help Files page about the **Table** command. The page begins by explaining the *syntax* of the **Table** command, that is, the different ways that we can use the command. Notice that six different forms are listed and so far we have discussed the third and fourth forms. The second form shows that we can not only omit the stepsize, but also the starting value of the counter. If we do this *Mathematica* will assume that the counter should start at 1. The first form of the command shows that we can even omit the name of the counter! This would be a fairly uncommon usage of **Table**, but sometimes it's just what we need. After showing the various forms of the **Table** command, the help page then gives a lot more information including examples and links to tutorials.

Sometimes we might remember the command we want but not quite remember how to use it. We can get a quick answer by typing a question mark followed by the name of the command directly into a *Mathematica* cell and evaluating that cell. Example 1.15.1 shows what happens. We get a description of the syntax and a hyperlink that will take us to the documentation shown in Fig. 1.3.

Example 1.15.1

In[62]:= (* getting information on a function *)

   ? Table

Table[*expr*, {i_{max}}] generates a list of i_{max} copies of *expr*.

Table[*expr*, {i, i_{max}}] generates

   a list of the values of *expr* when i runs from 1 to i_{max}.

Table[*expr*, {i, i_{min}, i_{max}}] starts with $i = i_{min}$.

Table[*expr*, {i, i_{min}, i_{max}, di}] uses steps di.

Table[*expr*, {i, {i_1, i_2, ...}}] uses the successive values i_1, i_2, ...

Table[*expr*, {i, i_{min}, i_{max}}, {j, j_{min}, j_{max}}, ...]

   gives a nested list. The list associated with i is outermost. ≫

Another way to reach the documentation page on a specific function, if the function name appears in our notebook, is to select the function name and then choose **Help ▶ Find Selected Function** from the menu bar.

If we can't remember, or don't know what function we need, select **Help ▶ Function Navigator** from the menu bar. This will bring up a catalog of functions organized in various categories. For example, under **Core Language ▶ List Manipulation ▶ Constructing Lists** we'll find **Table** (as well as 11 other functions). The **Function Navigator** is a great place to learn about new functions. Suppose we are working with lists and need to pick out a certain element from a list. By browsing the **Function Navigator** we can see what functions are available and we might find just what we need.

Finally, the **Virtual Book** is an excellent resource. It groups together all the guides and tutorials that are in the Help Files in an organized way. The Help Files are an indispensable source of information and the *Mathematica* user needs to learn how to use this valuable resource. We'll be offering a guide to the Help Files as we go.

1.16 Find Out More

In this chapter we have learned how to start *Mathematica*, type some basic commands into a notebook, and save our work. We have seen some of the syntactic features of *Mathematica* such as the fact that all built-in functions start with a capital letter, all functions use square brackets to enclose their arguments, and so on. We have also learned how to access the Help Files, an important source of information.

To find out more about getting started, we recommend you go through a couple of the *Mathematica* tutorials that can be found in the Help Files. The following tutorials and other information should be helpful.

- First Five Minutes with *Mathematica*—This is a very quick introduction that shows off a few functions that we'll be seeing in later chapters. Choose **Help ▶ Documentation Center** from the menu bar and look for the link to this tutorial in the lower right column of the page.
- The **Virtual Book**—Open the **Virtual Book** and start perusing it! Take a look at the **Introduction** and start reading the entries under **Getting Started**.
- The **Function Navigator**—*Mathematica* has over 2200 built-in functions. Open the **Function Navigator** and start looking around in it. You can also see an alphabetical list of all functions by choosing **Help ▶ Documentation Center** and then "Index of Functions" from the lower right column of the page. Try clicking on one of the functions. It will take you to the Help Files page for that function.
- Entering Expressions—Start reading **Notebooks and Documents ▶ Input and Output in Documents** in the **Virtual Book**. Don't skip this one!
- Building Up Calculations Overview—Read the **Building Up Calculations** section under **Core Language** in the **Virtual Book**.

As you browse through the documentation you will find other links that might be helpful. Lots of the Help File pages are not going to be of interest now, but will become useful as you learn more.

Quiz

1. Use *Mathematica* to compute $(\frac{1}{2} + \frac{1}{3})^3$ *exactly*.
2. Use *Mathematica* to compute $(\frac{1}{2} + \frac{1}{3})^3$ and represent the answer in decimal form.
3. It turns out that the numbers e^π and π^e are pretty close to each other. (Here e is the base of the natural logarithm and π is the ratio of the circumference of any circle to its diameter.) Without computing them it is not easy to decide which is bigger. Use *Mathematica* to find out which number is bigger.

4. The volume of a ball of radius r is given by $V = \frac{4}{3}\pi r^3$ and its surface area is given by $A = 4\pi r^2$. The radius of the earth is about 4000 miles. Use *Mathematica* to estimate the volume and surface area of the earth.

5. The volume of any cone is one-third the area of the base times the height, where the height is measured perpendicular to the base. The Great Pyramid at Cheops has a square base about 230 meters on a side and its height is about 147 meters high. Use *Mathematica* to compute the volume of the pyramid.

6. Use the **Table** function to make a list of the cubes of the first 10 integers.

7. The sine function is given by **Sin**. Use the **Table** function to make a table of $\sin x$ for every x from 0 to $\pi/2$ in increments of $\pi/20$. Have the entries in the table in decimal form.

8. The number e is defined as the limit of $(1 + 1/n)^n$ as n approaches infinity. Use the **Table** function to list the value of $(1 + 1/n)^n$ for $n = 10, 10^2, 10^3, \ldots, 10^6$. The numbers in this list are getting closer and closer to e. HINT: Replace n in the expression $(1 + 1/n)^n$ with 10^k and then let k go from 1 to 6.

9. The function $n!$, which is read "n factorial," is defined to be the product of all positive integers from 1 to n. Thus $3! = 1 \cdot 2 \cdot 3 = 6$, $5! = 1 \cdot 2 \cdot 3 \cdot 4 \cdot 5 = 120$, and the like. To compute $n!$ in *Mathematica* we may either type **Factorial[n]** or **n!**. Use the **Table** function to make a list of $n!$ for $n = 1$ to $n = 20$. (Note: It is conventional to *define* 0! to be 1. Try asking *Mathematica* to compute 0!)

CHAPTER 2

Two-Dimensional Graphics

Mathematica can be used to draw beautiful pictures that make it easy to visualize complicated curves, surfaces, data sets, or other shapes. In this chapter we'll focus on two-dimensional graphics. In Chap. 6 we'll introduce tools to display three-dimensional objects.

2.1 The Plot Function

One of the most fundamental and useful graphics tools is the **Plot** function which can be used to draw the graph of a function. Here is a simple example.

Example 2.1.1

```
In[74]:= (* using Plot to graph a function *)
    Plot[Sin[x], {x, 0, 2 Pi}]
```

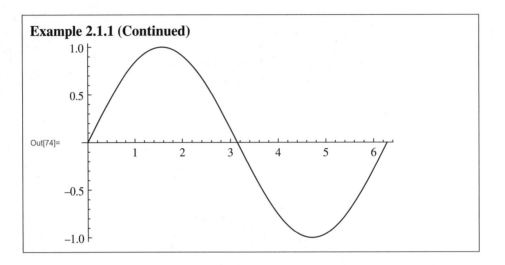

Example 2.1.1 (Continued)

Out[74]=

The **Plot** function takes two arguments. The first is the function that we want to plot which, in this case, is the sine function **Sin[x]**. The second argument is the list $\{$**x, 0, 2 Pi**$\}$ which tells *Mathematica* to graph $\sin x$ from $x = 0$ to $x = 2\pi$. In other words, this list indicates the *domain* of the function. The domain list $\{$**x, 0, 2 Pi**$\}$ is similar to the counter list $\{$**k, 1, 10**$\}$ that we might include as an argument to the **Table** function. In both cases we are naming a variable, x or k, and giving the minimum and maximum values that we want it to vary between. With the **Plot** function it is important that the variable we use in the function matches the one we use in the domain list. If we had entered **Plot[Sin[y], $\{$x, 0, 2 Pi$\}$]** it would not work because the two variables (x and y) do not match.

The first argument to the **Plot** does not need to be a single function. In fact it can be a list of functions, in which case *Mathematica* will superimpose the graphs of all the functions in the list. Let's superimpose the graphs of $\sin x$ and $x^2/10$. The next example will do this.

Example 2.1.2

```
In[76]:= (* plotting more than one graph *)
        Plot[{Sin[x], x^2 / 10}, {x, 0, 2 Pi}]
```

Example 2.1.2 (Continued)

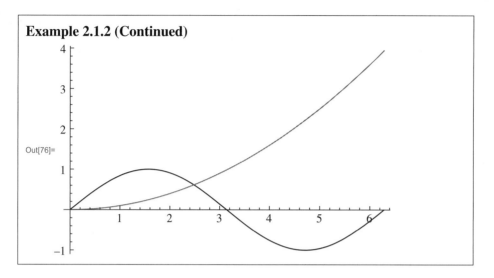

Out[76]=

2.2 Resizing Graphics

After using **Plot** we can resize the graphic using the mouse. First click anywhere in the graphic. This will display a *bounding box* surrounding the graphic as shown in Example 2.2.1. At the corners of the bounding box, and at the midpoints of its sides are small squares known as *handles*. If we drag one of the handles with the mouse the figure will change size. You have to try this yourself to see how it works!

Notice, however, that as you drag the bounding box, the *shape* of the figure will remain the same, that is, the ratio of the height to the width of the bounding box will remain the same. This ratio is called the *aspect ratio* of the figure. The aspect ratio remains constant as we drag the handles.

Example 2.2.1

In[1]:=   (* click graph to see the bounding box *)
          Plot [Sin [x], {x, 0, 2π}]

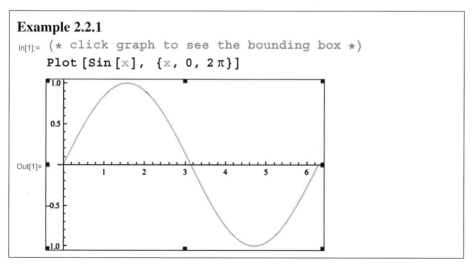

Out[1]=

If instead, we hold down the Shift key as we drag one of the handles then we can change the aspect ratio. Doing this to Example 2.2.1 allows us to make the figure wider and shorter as shown in Example 2.2.2.

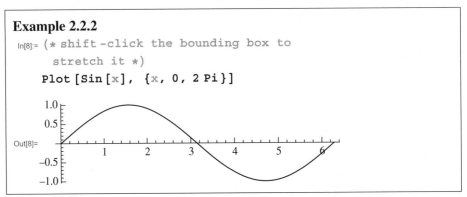

Example 2.2.2

In[8]:= (* shift-click the bounding box to
        stretch it *)
     Plot[Sin[x], {x, 0, 2 Pi}]

What happens if we drag the edge of the bounding box and not one of the handles? This will introduce *margins* around the figure with a new bounding box surrounding the original bounding box. By dragging the edge of the smaller box (not one of its handles), we may drag the smaller bounding box around inside the larger one to place it anywhere inside the larger box. If we hold down the Shift key as we do this, the inner box will automatically be centered in the outer box. To get rid of margins first drag the inner box to one corner of the outer box. Next resize the outer box to be as small as the inner one. Finally, we can crop a figure by holding down the Command key as we drag one of the handles.[1] You really need to try all of this yourself. Example 2.2.3 shows the outcome after first introducing margins and then cropping the inner box.

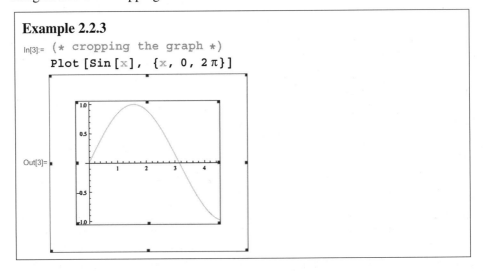

Example 2.2.3

In[3]:= (* cropping the graph *)
     Plot[Sin[x], {x, 0, 2π}]

---

[1] The Command key is a Mac feature—it does not exist on Windows or Linux machines.

2.3 Graphics Options

The function **Plot** can accept several really cool options for making our graphs look a lot better. Notice that in our very first plot of sin x that the scales on the horizontal and vertical axes are not the same. That is, 1 vertical unit appears to be quite a bit longer than 1 horizontal unit. If we want the scales on each axes to appear to be the same we need to change the aspect ratio of the graph. We could do this by selecting the graphic and resizing it, but then we would have to "eyeball" it. It would be difficult to get the two scales to be exactly the same length. On the other hand, we can use the **AspectRatio** option to control the aspect ratio exactly. The next example illustrates this.

In this example we see the basic syntax of a **Plot** option: the name of the option followed by the arrow followed by the value of the option. In this case **AspectRatio** is the name of the option and **Automatic** is the value we are giving to this option. (Note that the arrow is typed into the input cell by typing "->," a hyphen followed by the greater than sign. In most cases after typing ->, *Mathematica* will reset the arrow nicely. If not, try typing esc, ->, esc.)

Example 2.3.1

In[6]:= `(* controlling the aspect ratio *)`
`Plot [Sin [x], {x, 0, 2 Pi},`
`   AspectRatio → Automatic]`

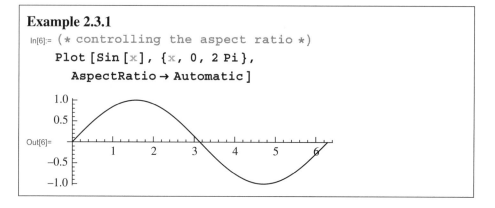

By giving the value **Automatic** to **AspectRatio** we have forced *Mathematica* to produce a picture that has just the right height and width so that the horizontal and vertical scales each have a unit that is the same length in the picture. Instead, we could assign any numerical value to the aspect ratio. For example, if we used **AspectRatio** $\rightarrow 1$ we will get a picture with equal height and width. But since our picture spans 2π units horizontally and 2 units vertically, we would not get a picture where 1 unit in each of the two directions would appear to be the same length. If we didn't want to use the **Automatic** value and instead wanted to provide the value ourselves, we would have to use an aspect ratio of $2/2\pi$ in this case. Using **Automatic** is nice because if we decide to change the domain to $0 \leq x \leq \pi$ for example, we don't have to recompute the aspect ratio in order to

Mathematica Demystified

maintain a horizontal and vertical unit of the same length; *Mathematica* will do it for us.

If we don't mention a graphics option in the **Plot** function, *Mathematica* will use preset *default* values for each of the options. In the case of **AspectRatio** the default value is 1 over the *Golden Ratio*, or $2/(1 + \sqrt{5})$. Thus the picture will come out as a *Golden Rectangle*, a rectangle that has a reputation as being the "most beautiful" rectangular shape![2]

We can control the exact region of the plot by using the option **PlotRange** $\rightarrow$ **{{xmin, xmax},{ymin, ymax}}**. Using this option will force the plot to extend horizontally from **xmin** to **xmax** and vertically from **ymin** to **ymax**. Here is an example.

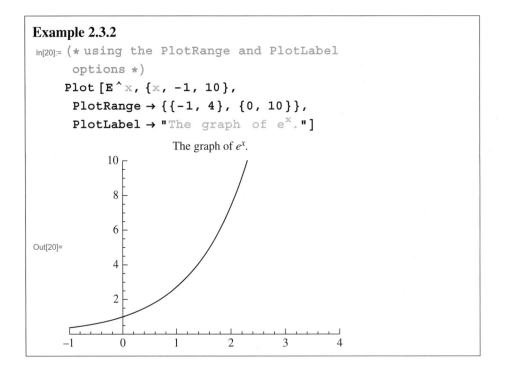

Example 2.3.2

```
In[20]:= (* using the PlotRange and PlotLabel
          options *)
       Plot [E^x, {x, -1, 10},
         PlotRange → {{-1, 4}, {0, 10}},
         PlotLabel → "The graph of e^x."]
```

The graph of e^x.

---

[2]A *Golden Rectangle* is a rectangle that has a very specific *shape*. In particular, it is not a square and if the rectangle is cut into a square and a smaller rectangle, the smaller rectangle still has the same shape as the larger one. The smaller rectangle is not as big as the original rectangle, but its dimensions still have the same ratio, which turns out to be $(1 + \sqrt{5})/2$, a quantity known as the *Golden Ratio*. This ratio was know to the ancient Greeks as the "extreme and mean ratio" and ever since then has generated more hoopla then almost anything else in mathematics; some of it interesting and some of it of dubious value. If you Google "golden ratio" you should get half a million hits or so.

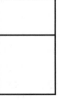

The plot extends horizontally from −1 to 4 and vertically from 0 to 10. Notice that we have also used the **PlotLabel** option to label the graph. This example shows that we can include multiple options—just separate them with commas. The BasicMathInput Palette was used here to type e^x rather than e^x in the label. We could also have typed esc, ee, esc, Ctrl+^, x, Ctrl+space. Finally, note that we have entered the function e^x as **E^x**. An alternative way to enter the exponential function is **Exp[x]**.

There are many times when we may want to override the default values that *Mathematica* chooses for a plot. For example, consider the graph of e^x in Example 2.3.3. Notice that the intersection of the two axes is not at the point (0, 0). If we want to force that to be true we can use the option **AxesOrigin** → **{0, 0}**. We'll leave this for you to try. (We could also force the axes to intersect at some point other than (0, 0). Try using **AxesOrigin** → **{0, 1}** for example.)

Example 2.3.3

```
(* The axes do not intersect at (0,0).
   We could change this with the AxesOrigin
   option. *)
Plot [E^x, {x, -1, 1}]
```

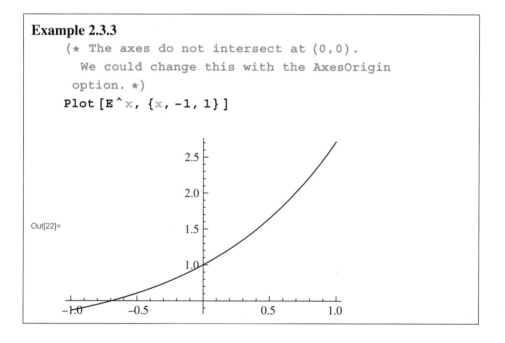

Out[22]=

Another useful graphics option is **PlotStyle** which allows us to change the way the actual graph of the function appears. In Example 2.3.4 we have used two different *attributes* for the *style*: **Dashed** and **Thickness**. The **Thickness** attribute takes an argument which, in this case, we have set to .005. This means that the thickness of the curve will be .005 times as wide as the figure itself. Thus if we resize the figure by clicking on it and then dragging the bounding box, the curve will get thicker as the figure gets wider.

Example 2.3.4
```
(* using the PlotStyle option *)
Plot[4 - x^2, {x, -2, 2},
  PlotStyle → {Dashed, Thickness[.005]}]
```

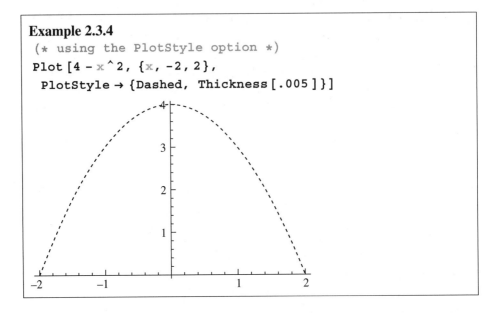

When graphing more than one curve at once it is useful to draw the different curves in different colors, thicknesses, and so on. Since this book is not printed in color, let's graph two functions and have each one have a different thickness. Example 2.3.5 shows how to do this. Since **PlotStyle** has two attributes, each a different thickness, the first attribute is applied to the first graph, and the second attribute to the second graph. We could do this with three or four or more graphs and three or four or more attributes. If the number of attributes exceeds the number of graphs, then the last couple of attributes will simply not be used. If on the other hand, the number of graphs exceeds the number of attributes, then when *Mathematica* runs out of attributes it will simply return to the beginning of the attribute list and start over.

Sometimes we may want to give several attributes to a single graph when we are graphing more than one graph. Suppose we want to graph three graphs and

Example 2.3.5
```
(* using Thickness with PlotStyle *)
Plot[{Sin[x], x^2/10}, {x, 0, 2 Pi},
  PlotStyle → {Thickness[.02],
    Thickness[.005]}]
```

Example 2.3.5 (Continued)

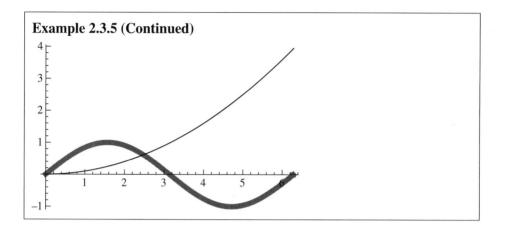

make each a different shade of gray and each a different thickness. Consider Example 2.3.6.

Here we have graphed three function: $\sin x$, $x - x^3/3!$ and $x - x^3/3! + x^5/5!$. If you have had calculus you might recognize the second and third functions as the Taylor polynomials of degree 3 and 5 for $\sin x$. The two polynomials provide pretty

Example 2.3.6

```
(* using GrayLevel with PlotStyle *)
Plot [{Sin [x], x - x^3 / 3!,
    x - x^3 / 3! + x^5 / 5!}, {x, -2 Pi, 2 Pi},
  PlotStyle → {{GrayLevel [0], Thickness [.006 ]},
    {GrayLevel [.2], Thickness [.004 ]},
    {GrayLevel [.5], Thickness [.003 ]}}]
```

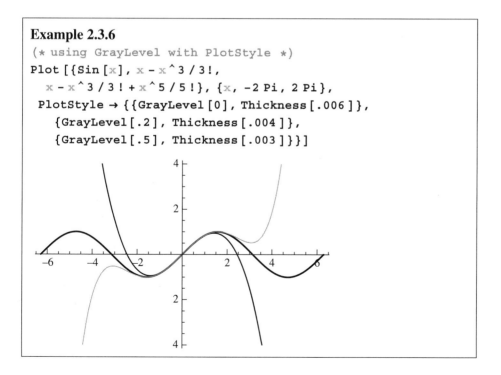

good approximations to $\sin x$ for small values of x, so near the origin, the graphs appear to overlap. Away from the origin we see the graphs start to diverge.

We have given a list of three sets of attributes to **PlotStyle**, with each set being a pair: **GrayLevel** and **Thickness**. The first pair is {**GrayLevel[0], Thickness[.006]**}. This causes the first function, $\sin x$, to be graphed in solid black with a thickness of .006. (A gray level of 0 is black; a gray level of 1 is white.) The second pair specifies a gray level of .2 and a thickness of .004. The second function is the cubic polynomial $x - x^3/3!$. Thus it is drawn slightly lighter and slightly thinner than the sine curve. Finally, the third function, the fifth degree polynomial, is graphed using the lightest shade of gray and the smallest thickness.

Another nice thing to do with a graph is to label the axes, which we can do using the **Frame** and **FrameLabel** options. Suppose we want to illustrate the relationship between the temperature, in degrees Fahrenheit, and the rate at which the snowy tree cricket chirps. According to an article that appeared in *Outside* magazine in June, 1995, counting the number of chirps in a 13 second period and then adding 40 gives a good approximation to the temperature. In Example 2.3.7 we have labeled the horizontal axis "Chirps in 13 seconds" and the vertical axis "Degrees Fahrenheit." Since we might want to actually read the temperature off of the graph for a given chirp rate, we have also used the **GridLines** option. **LabelStyle** is yet another option and we have used it to make all the labeling bold. This example again illustrates that we may use as many options as we want—we just need to separate them with commas.

Example 2.3.7

```
(* using GridLines,Frame,
and FrameLabel *)
Plot [x + 40, {x, 10, 50},
 PlotLabel →
  "Temperature vs. Chirp Rate for Snowy
    Tree Cricket ",
 GridLines → Automatic , Frame → True,
 FrameLabel → {"Chirps in 13 seconds",
   "Degrees Fahrenheit "},
 LabelStyle → Bold ]
```

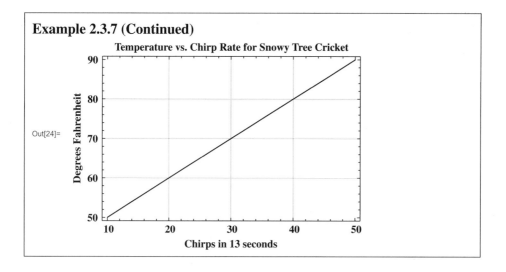

Example 2.3.7 (Continued)

Out[24]=

Before closing this section we mention one more useful option, **Ticks**. In each of our plots *Mathematica* has placed little tick marks on the axes and labeled them with values. We can control this ourselves by using **Ticks**. Example 2.3.8 shows how.

Notice that we have specified two lists of numbers for the **Ticks** option. The first is the list {-2 π, -3 π/2, -π, -π/2, π/2, π, 3 π/2, 2 π}, which we generate with the **Table** function, and the second is {-1, -.5, .5, 1}. We could have used a **Table** function to generate the second list of ticks too, but in this case it seems easier to

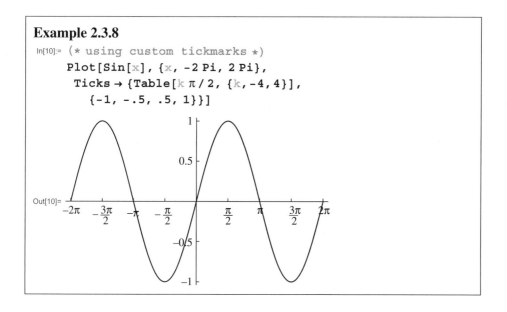

Example 2.3.8

```
In[10]:= (* using custom tickmarks *)
     Plot[Sin[x], {x, -2 Pi, 2 Pi},
       Ticks → {Table[k π / 2, {k,-4,4}],
         {-1, -.5, .5, 1}}]
```

Out[10]=

Mathematica Demystified

just type them out. *Mathematica* uses the first list as the locations of tick marks (and tick labels) on the horizontal axis and the second list as the location for tick marks on the vertical axis.

We have only highlighted a few of the many options that are available with the **Plot** function. To see a list of all such options enter **Options[Plot]**. We can also use the **Options** function to see what options are available for **Plot3d**, **ParametricPlot**, and **ContourPlot**, to name just a few of the other plotting functions that we will be learning about.

2.4 Drawing Tools and the Graphic Inspector

After drawing a graphic using **Plot**, we can add additional features to the graph using the drawing tools. Choosing **Graphics ▸ Drawing Tools** from the menu bar will bring up the 2D Drawing palette shown in Fig. 2.1.

The palette contains a number of *tools*, with the upper-right tool currently selected. (It has a colored box around it to indicate that it is selected.) This tool is the *Select/Move/Resize* tool. Suppose we want to insert some text into our figure, perhaps labeling something in the figure. To do this, first select the figure by clicking anywhere in it. Next select the *Text* tool from the 2D Drawing palette. Now position the mouse in the figure at the place where you want to insert the text, click the mouse, and start typing the text. After typing the text, press Shift+Return. Doing this we can label the sine curve with its equation as shown in Example 2.4.1.

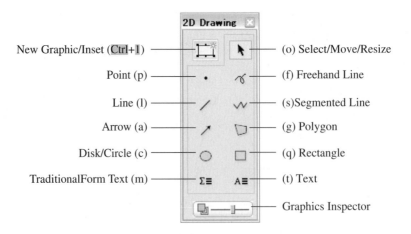

Figure 2.1 The 2D Drawing palette.

Example 2.4.1

In[4]:= (* adding text with the Drawing Tools *)

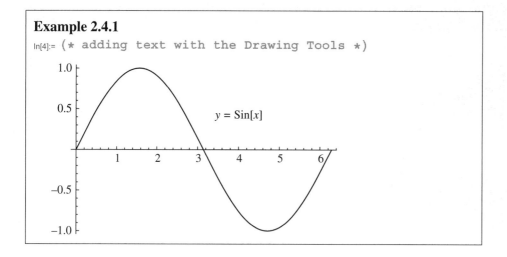

If we want to edit the text, first select it. We can do this by double-clicking on it. After it is selected it will be surrounded by a bounding box. We could change to a bigger font, for example, by going to the **Format ▶ Size** menu and then choosing a larger point size. Or, after selecting the text, we may want to drag it to a different location.

By choosing other tools it is possible to draw lines, arrows, circles, rectangles, and other objects in the figure. You should experiment with the different tools to see what they can do. There is a also a nice tutorial in the Help Files that we will direct you to at the end of this chapter.

At the very bottom of the 2D Drawing palette is a button for the "Graphics Inspector." Clicking this button will bring up the Graphics Inspector palette shown in Fig. 2.2. We can use this to change, for example, the color or thickness of a curve. First double click on the curve to select it. Next adjust the color or thickness using the palette. Using this method has almost exactly the same effect as using the plot style options **Thickness** or **GrayLevel**. The only difference is that using the Graphics Tools creates a second output cell with the altered drawing. If we reenter the input cell, the original graphic (without the added features) will be regenerated. The altered drawing with the added features will remain from before.

The drawing tools can be used to do some things that would be incredibly difficult to accomplish with just the **Plot** options. On the other hand, some effects created with **Plot** options cannot be done afterward with the graphics tools. Being able to use both gives you the most control over your graphics.

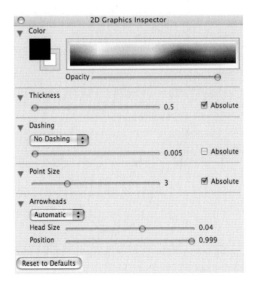

Figure 2.2 The Graphics Inspector palette.

2.5 Using Epilog

Another way to add labels or other graphics objects to a plot, rather than using the Drawing Tools, is to use **Epilog**. With the **Epilog** option we can have *Mathematica* add additional *graphics objects* to a plot after the plot is rendered. (There is a **Prolog** too that adds objects before the plot is rendered.) For example, we might want to label different features within a plot. In Example 2.5.1, we have used **Epilog** to add **Text** *objects* to the plot after the graphs are plotted.

The **Text** function takes two arguments. The first is the text string that we want to insert in the plot. The second is the coordinates of the point where the text should be centered. The coordinates we gave for "Factory B" are {5, 10} so the center of this text string is placed at the point (5, 10) in the plot. For the "Factory A" label we have used the **Style** function which allowed us to alter the font size by using the **FontSize** option. Of course, in this example, we could have easily put in the labels by using the Drawing Tools, but sometimes using **Epilog** is a better choice. We'll see this in Chaps. 7 and 10 where **Epilog** is used to add graphical elements that we could not possibly add with Drawing Tools.

Example 2.5.1

In[33]:=

```
(* using Epilog to add labels *)
Plot[{20 Exp[.01 t], 12 Exp[.03 t]},
 {t, 0, 40},
 AxesOrigin → {0, 0},
 Epilog →
  {Text[Style["Factory A", FontSize → 18],
    {5, 25}],
   Text["Factory B", {5, 10}]}
]
```

Out[33]=

2.6 Mouseover Effects—Tooltip

An alternative to providing permanent labels in a plot is to use **Tooltip** to provide "mouseover" labels. These are labels that will appear only when the mouse is moved over a certain feature in the plot. In Example 2.6.1 we provide labels for each curve, but these labels only appear when the mouse pointer is moved over the curve. In this example, the label for the "Factory B" curve has "popped up" because the cursor was moved over the curve.

Here we have modified each of the two function that we want to graph with the **Tooltip** function. Whereas before we would have simply had the function **20**

Example 2.6.1

In[39]:=

```
(* using Tooltip for mouseover effects *)
Plot[{Tooltip[20Exp[.01 t], "Factory A"],
   Tooltip[12Exp[.03 t], "Factory B"]},
   {t, 0, 40}, AxesOrigin → {0, 0}]
```

Out[39]=

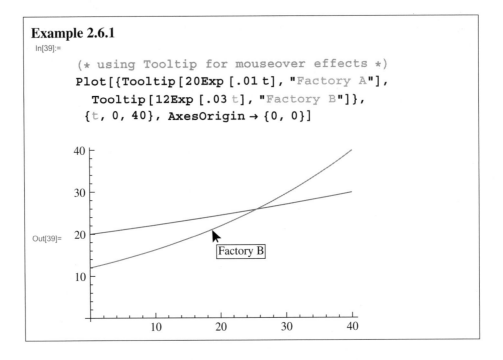

Exp[.01 t] we now have **Tooltip[20 Exp[.01 t], "Factory A"]**. The **Tooltip** function takes two arguments. The first is whatever object we want to label when the mouse is placed over it and the second argument is the label. So, in this example, when the mouse is moved over the graph of $20e^{.01t}$ a label of "Factory A" will appear. You need to try this yourself! Try it!

2.7 Animation—Using Manipulate

One of the really great improvements in *Mathematica* versions 6 and 7 is the capability of *interacting* with a *Mathematica* notebook. We have already seen one example of this with the mouseover effects made possible with the **Tooltip** function. The **Manipulate** function also allows the user to interact with graphics (and other things too) in a really neat way.

Suppose we want to understand the cubic polynomial $x^3 + ax + 1$, where a is some constant. In particular, maybe we want to know how the value of a affects the graph. We could start picking various values for a and then plot the graph for each value that we have chosen, but a better way is to use **Manipulate**. Consider Example 2.7.1. We have taken the **Manipulate** function and *wrapped* it around the

Plot function. In the **Plot** function we have plotted the function $x^2 + ax + 4$ from $x = -8$ to $x = 8$. We also have included a couple of **Plot** options: **PlotRange** and **AspectRatio**. The variable of the cubic polynomial is x. But a is a variable too! We call it a *parameter*. As we change the parameter we get a different function. What we really have is a whole *family* of functions, one for each value of a. The **Manipulate** function allows us to dynamically change the parameter a and watch the graph of the corresponding cubic polynomial change! This book, of course, is totally static! You *have* to try this out for yourself with *Mathematica*!

Example 2.7.1

In[41]:=

```
(* using Manipulate for animation *)
Manipulate[
 Plot[x^3 + a x + 4, {x, -8, 8},
  PlotRange → {{-8, 8}, {-100, 100}},
  AspectRatio → 1],
 {a, -10, 10}
]
```

Out[41]=

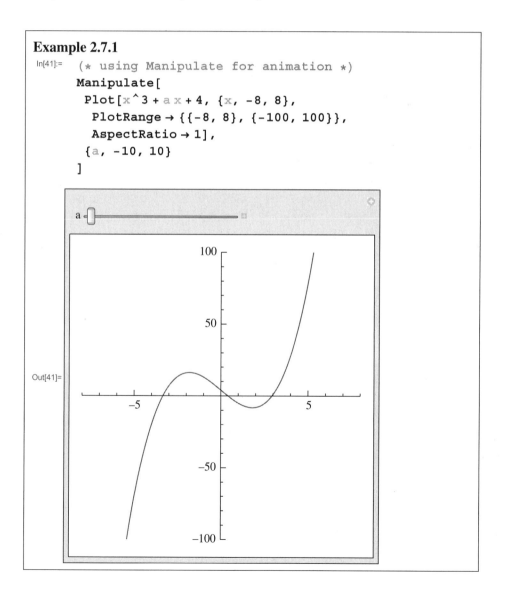

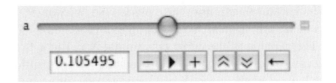

Figure 2.3 The expanded slider control.

Notice that the output cell contains the graph of the function together with a slider control labeled "a". As we drag the slider control with the mouse we change the value of *a* and thus change the function and hence its graph. This is awesome! The **Manipulate** function takes two arguments. The first is the object we wish to *manipulate*, in this case the **Plot** which contains the parameter *a*. (We'll see later that we can manipulate other things too, not just plots.) The second argument is the parameter control list {**a, -10, 10**}.

Notice that just to the right of the slider control is a small box with a plus sign in it. If we click this, the slider control expands to show more information and appears as shown in Fig. 2.3. The value of the parameter *a* is now displayed as we move the slider. To the right of this display field, we have six buttons. Clicking the plus and minus signs will increase or decrease the parameter by one. (But note that by using the slider you can obtain noninteger values of *a*.) Clicking the "play" button between them will cause the parameter to change continuously thus giving us an animation! Clicking it again will pause the animation. The animation can be sped up, slowed down, or reversed by using the next three buttons.

There are quite a few additional optional features that can be used with **Manipulate**. We'll introduce more of them later in the book. In the "Find Out More" section at the end of this chapter we'll direct you to a nice *Mathematica* tutorial on **Manipulate**.

2.8 Plotting Points with ListPlot

Instead of plotting the graph of a function, we sometimes need to plot a set of points—perhaps data that was collected in some experiment. For example, returning to the snowy tree cricket, suppose we have collected the data in Table 2.8.1 by listening to an actual cricket, measuring the temperature with a thermometer and also counting the number of chirps in a minute.

Table 2.8.1   Chirp Data Recorded for a Snowy Tree Cricket

Chirps/minute	Temperature
30	46.5
34	47.2
35	47.6
37	48.2
42	49.2
49	50.8
50	50.8
55	52.0
61	53.3

The function **ListPlot** can now be used to plot this data. Example 2.8.1 shows how to do this. **ListPlot** takes many of the same options as **Plot** and we have also used the **PlotRange, AxesOrigin**, and **PlotStyle** with **PointSize** options.

Example 2.8.1

```
(* plotting datapoints with ListPlot *)
chirpdata = {{30, 46.5}, {34, 47.2}, {35, 47.6},
    {37, 48.2}, {42, 49.2}, {49, 50.8},
    {50, 50.8}, {55, 52.0}, {61, 53.3}};
dataPlot = ListPlot[chirpdata,
  PlotRange -> {{25, 65}, {45, 55}},
  AxesOrigin -> {25, 45},
  PlotStyle → PointSize[.02]
  ]
```

Out[17]=

Table 2.8.2 Population (in millions) for United States, India, and China

Year	USA	India	China
1970	210.111	549.312	815.999
1975	220.165	613.767	911.658
1980	230.917	688.575	981.072
1985	243.063	771.121	1047.59
1990	256.098	860.195	1128.67
1995	270.245	954.282	1192.37
2000	284.857	1046.24	1247.69
2005	299.846	1134.4	1290.21

Note that we first enter the data into a list called **chirpdata**. We should think of this list as a list of points, where, of course, each point is itself a list of two coordinates: number of chirps per minute and temperature. Next we use **ListPlot** to plot the points. Notice that we could have just entered the data into the first argument of **ListPlot** and not have taken the trouble to first name it **chirpdata**. While this would have worked just fine, it is nicer to name the data and then use the name as the argument to the plot function. Separating it this way makes it easier to see what is going on.

A nice option to **ListPlot** is the **Joined** option. If we add **Joined→True** to the list of options, consecutive points in the plot will be joined together with line segments. Try it!

ListPlot can be used to plot multiple lists of data simultaneously just as **Plot** can be used to graph more than one function at a time. Consider, for example, the population data, in millions of people, from three different countries over a 35-year period given in Table 2.8.2. In Chap. 9, we'll see how to use *Mathematica* to import this data from the **CountryData** data set maintained by Wolfram Research.

Let's use **ListPlot** to plot all the data at once. First we need to organize the data into three lists, one for each country, which we call **usData, indiaData**, and **chinaData**. Then we can use **ListPlot** to plot the data. The first argument to **ListPlot** is the list {**usData, indiaData, chinaData**}. The remaining arguments are all options. Notice that we have used the option **PlotMarkers→Automatic**. This causes different symbols (circles, squares, and diamonds) to be plotted for each of the data sets.

Example 2.8.2

```
In[67]:=  (* using ListPlot to plot datapoints *)
        usData = {{1970, 210.111}, {1975, 220.165},
          {1980, 230.917}, {1985, 243.063},
          {1990, 256.097}, {1995, 270.244},
          {2000, 284.857}, {2005, 299.846}};
        chinaData = {{1970, 815.999}, {1975, 911.657},
          {1980, 981.072}, {1985, 1047.592},
          {1990, 1128.667}, {1995, 1192.374},
          {2000, 1247.685}, {2005, 1290.208}};
        indiaData = {{1970, 549.312}, {1975, 613.767},
          {1980, 688.575}, {1985, 771.120},
          {1990, 860.195}, {1995, 954.281},
          {2000, 1046.235}, {2005, 1134.403}};
        ListPlot[{usData, indiaData, chinaData},
         AxesOrigin → {1969, 0},
         PlotRange -> {{1969, 2006}, {0, 1300}},
         PlotMarkers → Automatic, Joined → True
        ]
```

Out[70]=

2.9 Curve Fitting

With data sets like those shown in Examples 2.8.1 and 2.8.2 we often want to find the *best fitting curve* that approximates the data. *Mathematica* has a nice function called **Fit** that will do just this. In Example 2.9.1, we use **Fit** to find the best fitting line to the snowy tree cricket data. **Fit** takes three arguments. The first is the list of data points. Next is a list of functions, in this case the constant function 1 and the linear function x. The final argument is the independent variable. Using the method of least squares approximation, **Fit** will find the best *linear combination* of these two functions that fits the data. A linear combination of any set of functions is a sum of constant multiples of each function. Thus a linear combination of 1 and x is a function of the form $a + bx$, where a and b are constants. But this is just the most general possible linear function! Hence, in this case, **Fit** is finding the best fitting line.

In the second part of the example, we use **Show** to combine, or superimpose, two different plots. Notice that we named the plot of Example 2.8.1 "dataPlot" and that we enter that name as an argument to **Show** here. We'll be talking more about **Show** in Chap. 6, so hold on until then. But, the point is that the linear approximation clearly fits the data points well.

Example 2.9.1

```
In[30]:= (* finding the best line to fit the data *)
      Fit[chirpdata, {1, x}, x]

Out[30]= 39.8931 + 0.220259 x

In[34]:= (* using Show to combine plots of the
         data and line *)
      Show[
       dataPlot,
       Plot[39.8931 + 0.220259 x, {x, 30, 61}]
      ]
```

Out[34]=

Returning to the population data of Example 2.8.2, let's fit a curve to India's population data. From Example 2.8.2, we can see that the rate at which the population is growing seems to be increasing. So attempting to fit the data with a line is not appropriate. In Example 2.9.2 we find the best fitting parabola, or second degree polynomial, by finding the best linear combination of 1, x, and x^2. Again, we plot the data and the curve separately and then combine them using **Show**. (We use semicolons to suppress the first two plots.)

Example 2.9.2

```
In[77]:= (* fitting a parabola to the India
        population data *)
     Fit[indiaData, {1, x, x^2}, x]
```

Out[77]= $306\,406. - 324.522\,x + 0.08592\,x^2$

```
In[87]:= (* plotting the data and the parabola *)
     curvePlot = Plot[
        306406. - 324.522 x + 0.08592 x^2,
        {x, 1970, 2005}
        ];
     dataPlot = ListPlot[
        indiaData,
        PlotRange -> {{1970, 2005}, {0, 1300}}
        ];
     Show[dataPlot, curvePlot]
```

Out[89]=

2.10 Plotting in Polar Coordinates

All of the graphs we have plotted so far have been with respect to the Cartesian coordinate system, which is just one way to associate coordinates to points in the plane. Another way is given by the *polar* coordinate system which still uses just two coordinates but this time the coordinates are a distance, or radius, r and an angle θ. Just as with Cartesian coordinates, the system starts with a point of reference with respect to which the coordinates can then be used to locate points. In this case the reference is a *polar ray* which originates at the *pole*. Once the pole and the polar ray are given, then a point with coordinates (r, θ) is located by first moving r units away from the pole along the polar ray and then rotating counterclockwise through an angle of θ. It is convenient to allow the radius r to be negative, in which case we first start by marking off the distance in the *opposite* direction as the polar ray. We may also allow the angle θ to be negative, in which case we rotate clockwise. *Mathematica* has two functions, **PolarPlot** and **ListPolarPlot** that are the analogues of **Plot** and **ListPlot** for polar coordinates. Let's look at some examples.

Example 2.10.1

```
In[1]:= (* plotting points in polar coordinates *)
    ListPolarPlot[
     {{Pi / 4, 1}, {Pi / 2, .5}, {0, -2.5},
      {- 3 Pi / 4, 1}},
     PlotStyle→ {PointSize[.02]},
     AspectRatio→ Automatic]
```

Out[1]=

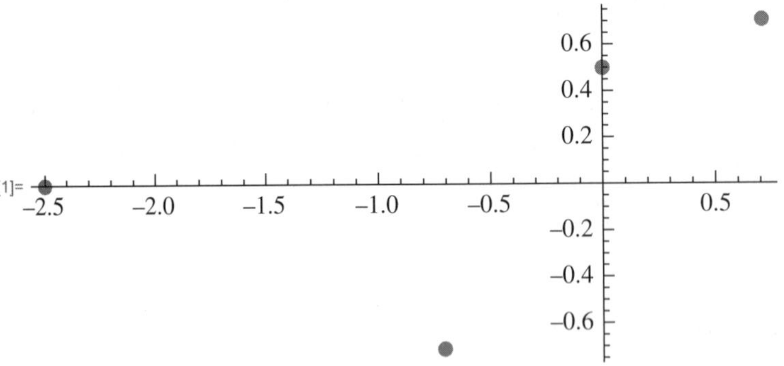

In Example 2.10.1 we simply use **ListPolarPlot** to plot four points. *Mathematica* uses the standard convention of superimposing the polar and Cartesian coordinate systems so that the polar ray (in the polar system) coincides with the positive horizontal axis (in the Cartesian system). When using **ListPolarPlot** the points are

given with the angular coordinate first and the radial coordinate second. Thus the first point in the list, $(\pi/4, 1)$, appears in the upper-right part of the plot. The point is 1 unit from the pole and rotated $\pi/4$ radians away from the polar axis in the counterclockwise direction. Note that the third point, which has an angle of zero and a radius of -2.5, is marked off on the *negative* horizontal axis because the radius is negative.

In Example 2.10.2, we give a second example of plotting a list of points using **ListPolarPlot**. In this case, the points are equally spaced at angles of $\pi/100$ but have radii that depend on the angle.

Example 2.10.2

```
In[96]:= (* plotting a set of points in polar
          coordinates *)
       ListPolarPlot[
       Table[{θ, Sin[2 θ]}, {θ, 0, 2π, π / 100}]
       ]
```

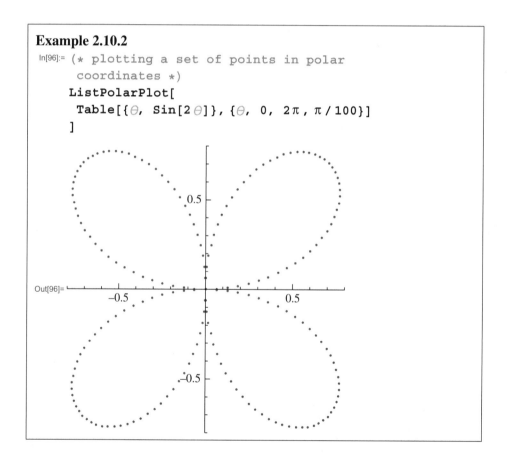

Of course, these points lie on the curve whose polar equation is given by $r = \sin(2\theta)$. To graph this curve we can use **PolarPlot** as seen in Example 2.10.3.

Just like **Plot**, we can plot multiple functions at once with **PolarPlot** as well as use all the usual options. In Example 2.10.4, we graph two functions at once and make each curve a different thickness.

Example 2.10.3

In[98]:= (* plotting a function in polar coordinates *)
PolarPlot[Sin[2 θ], {θ, 0, 2 π}]

Out[98]=

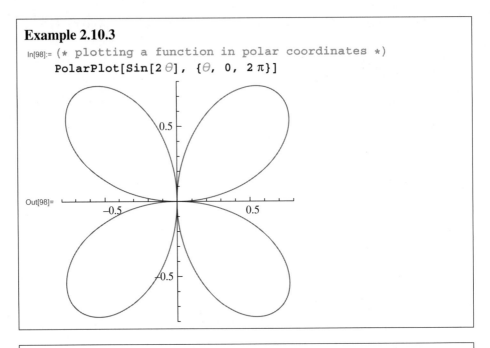

Example 2.10.4

In[106]:= (* plotting multiple functions *)
PolarPlot[
 {Sin[3 θ], Cos[θ] + Sin[θ]},
 {θ, 0, 2 π},
 PlotStyle → {Thickness[.01], Thickness[.02]}
]

Out[106]=

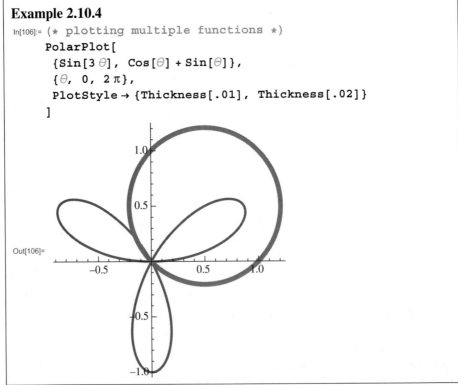

2.11 Parametric Plots

It is often convenient to describe a curve parametrically in terms of some *parameter*. For example, suppose a particle is moving in the plane and at time t is located at the point $(x(t), y(t))$. (Here we are using Cartesian coordinates.) We say that the curve traced out by the particle is described *parametrically by the parameter t*. A nice family of curves of this form are the *Lissajous* curves, where $x(t)$ and $y(t)$ are given by

$$x(t) = \cos(n_x t + \phi_x)$$
$$y(t) = \cos(n_y t + \phi_y).$$

The integers n_x and n_y are called the *frequencies*, and the real numbers ϕ_x and ϕ_y are called the *phase shifts*. In Example 2.11.1, we plot this curve using the **ParametricPlot** function.

Example 2.11.1

In[25]:= (* plotting a Lissajous curve *)
        ParametricPlot[{Cos[3 t], Cos[5 t + Pi / 5]},
         {t, 0, 2 Pi}]

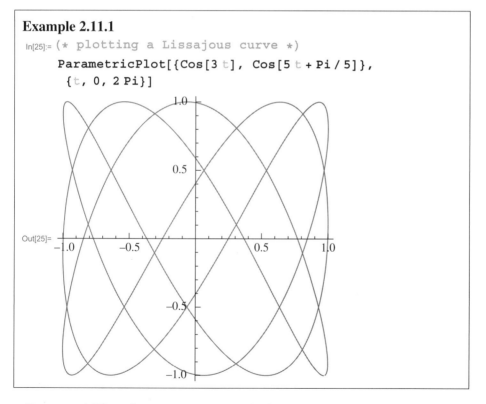

ParametricPlot takes two arguments. The first is the list of coordinate functions, each in term of the parameter, and the second argument is the familiar domain list for the parameter. As usual, all sorts of options can be added.

The Lissajous curves are closely related to *billiard trajectories*. Imagine rolling a billiard ball on a square billiard table. It will travel in a straight line until it bounces off a rail, then continue in a straight line until hitting the next rail, and so on. Suppose the table is one unit long on each side. Let's introduce Cartesian coordinates so that the corners of the table are at $(0, 0)$, $(1, 0)$, $(1, 1)$, and $(0, 1)$. If the ball starts at the point (h, k) and v_x and v_y are the velocities of the ball in the x and y directions (and let's suppose there is no friction, so that these stay constant and the ball rolls forever), then the position of the ball at time t is given by

$$(1 - |\text{Mod}[v_x t + h, 2] - 1|, 1 - |\text{Mod}[v_y t + k, 2] - 1|)^3$$

Let's use **ParametricPlot** to draw the trajectory and **Manipulate** to vary the initial position (h, k) as well as the initial velocities v_x and v_y. The next example does this. The input cell appears in Example 2.11.2 and the output cell appears in Example 2.11.3.

Example 2.11.2

```
(* billiard trajectories on a square table *)
Manipulate[
  (* plot the path parametrically *)
  ParametricPlot[
    {1 - Abs[Mod[hspeed t + pt[[1]], 2] - 1],
     1 - Abs[Mod[vspeed t + pt[[2]], 2] - 1]},
    {t, 0, pathLength},
    (* plot options *)
    PlotRange → {{-.01, 1.01}, {-.01, 1.01}},
    Axes → False,
    PerformanceGoal → "Quality",
    (* use Epilog to draw billiard table *)
    Epilog → {
      Thickness[.01],
      Line[
        {{0, 0}, {0, 1}, {1, 1}, {1, 0}, {0, 0}}
      ]
    }
  ],
```

---

[3] It is an interesting exercise to derive these formulas for the coordinates. Think of tessellating the plane with an infinite number of billiard tables by starting with the original table and then reflecting it across its sides. In this "unfolded" world, a billiard trajectory is a straight line given by $(v_x t + h, v_y t + k)$. Reducing the coordinates modulo 2 will translate the ball into one of the four tables in $[0, 2] \times [0, 2]$, where the absolute value function can now be used to accomplish the reflection.

Example 2.11.2 (Continued)
```
(* controllers *)
Style["Speed", 12, Bold],
{{hspeed, .2, "Horizontal"}, -1, 1},
{{vspeed, .5, "Vertical"}, -1, 1},
Delimiter,
Style["Initial Position", 12, Bold],
{{pt, {.5, .5}, ""}, {0, 0}, {1, 1}},
Delimiter,
Style["Length", 12 , Bold],
{{pathLength, 5, ""}, .01, 100}
]
```

The instructions inside the **Manipulate** function consist of the **ParametricPlot** and then the controllers for **Manipulate**. Let's see how these work. The first two arguments to the **ParametricPlot** function are the coordinates of the point described in terms of the parameter t, and then the domain list for the parameter. Notice that the upper limit for t is **pathLength** which is a parameter that will be a control for **Manipulate**.

After the domain list we have four options. The first is the familiar **PlotRange**, but notice that we have set the range from just a little under zero to just a little over 1 in each direction. Looking ahead to the option **Epilog** that is used to plot the edges of the billiard table, we see that by making the plot range a little oversize we leave room for the thicker edges of the table. We'll explain the **Line** function in the next section. We use **Axes→False** to turn off the axes. Finally, the option **Performance Goal→ "Quality"** is needed to make the plot look good as the sliders are being moved. Look this option up in the Help Files to learn more about it.

Moving on to the controllers, we have introduced some neat formatting instructions as well as a two-dimensional slider to control both coordinates of the initial position simultaneously. If you haven't taken the time to look up **Manipulate** in the Help Files yet, you are really missing out! There are several more really cool features that we have yet to discuss. Notice the use of **Delimiter** to draw the lines separating the controllers. We have also used text strings to label the controllers, and further enhanced these strings by using the **Style** function which allows us to

Mathematica Demystified

Example 2.11.3

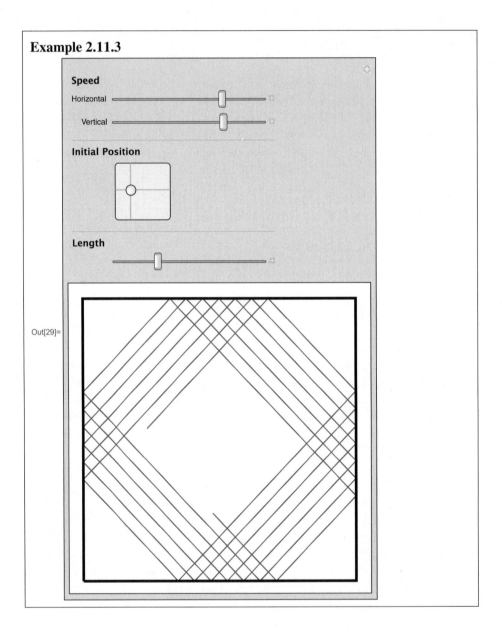

change the font face and size. Since we used text strings to label the controllers, we did not want to use the regular control labels for either of the parameters **pt** or **pathLength**, so we have included empty strings "" for each of those. The **Manipulate** Help page is quite good, so take a look there if this still seems a little mysterious.

When you play with this example, make sure you take the **pathLength** up to its maximum value. With a really long path, and just the right initial velocities, the trajectory can start to fill up the entire table. Very slight changes in the velocities can now make a really big difference in the outcome. A really nice feature of *Mathematica* is that by holding down the Option key (Macintosh) or Alt key (Windows), the action of the slider will become more sensitive to the movement of the mouse. In fact, the motion of the slider will be slowed down by a factor of 20 compared with the motion of the mouse. If the Shift or Control keys are also held down (in addition to the Option/Alt key) another factor of 20 for each additional key can be achieved. Thus, very accurate control of the sliders can be obtained.

2.12 Drawing Shapes

Instead of drawing the graphs of functions, or plotting points, we can use *Mathematica* to plot *graphics primitives* such as lines, circles, disks, rectangles, or polygons and, of course, we may use various options to change their appearance.

Suppose, for example, we want to draw a rectangle with vertices at $(0, 0)$, $(2, 0)$, $(2, 3)$, and $(0, 3)$. Example 2.12.1 will do this.

Here we have used **Rectangle[(0, 0),(2, 3)]** to create a rectangle with lower-left corner at $(0, 0)$ and upper-right corner at $(2, 3)$. In general, the syntax for the **Rectangle** function is **Rectangle[(x_{min}, y_{min}), (x_{max}, y_{max})]**. Next we used the **Graphics** function to display the *graphics object* that is given by the **Rectangle** command.

Example 2.12.1

```
In[34]:= (* drawing rectangles *)
     Graphics[Rectangle[{0, 0}, {2, 3}],
       Axes → True, PlotRange → {{-1, 3}, {-1, 4}}]
```

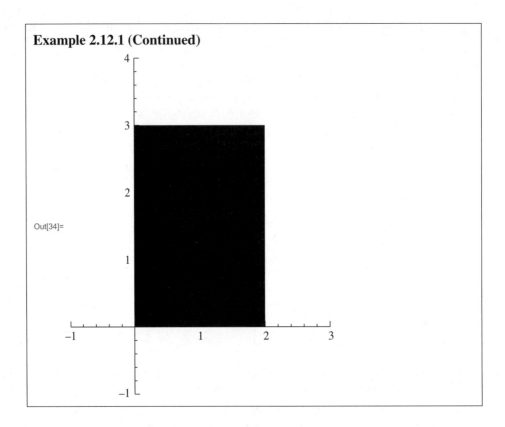

Example 2.12.1 (Continued)

Out[34]=

We could have simply used **Graphics[Rectangle[(0, 0),(2, 3)]]** but instead have included two options. Without the options we simply would have produced a big black rectangle and we would not have been able to tell that in fact its corners were where we wanted them to be. But by adding the options of **Axes** and **PlotRange** we can see that the rectangle is indeed located where we want. The syntax for the **Graphics** function is **Graphics[*primitives, options*]**. In this example we have only one primitive, namely, the single rectangle. If we have more primitives we simply provide them in a list.

In Example 2.12.2, we plot a polygon, circle, disk, and line segment. Notice that we do not use any options—the only argument to the **Graphics** function is the list of the four primitive objects. The syntax for each of these is as follows. For the polygon, we provide the list of vertices as we travel around the perimeter of the polygon. For the circle, we provide the center and then the radius, which is the same as for the disk. The difference between the circle and the disk is that the disk is filled in. Finally, for a line segment we provide a list of points. In this case we have provided only two points and they act as the endpoints of the line segment. But if we had provided a list of more than two points, *Mathematica* would draw a sequence of line segments connecting each of the points in succession. (This is known as a *polygonal path*.) We did this in Example 2.11.2 when we drew the edge of the billiard table using **Epilog**.

Example 2.12.2

In[35]:= `(* drawing various shapes *)`

```
Graphics[
 {Polygon[{{0, 0}, {1, 2}, {3, 4}, {6, 2},
    {0, 0}}],
  Circle[{4, 5}, 1],
  Disk[{1, 4}, .5],
  Line[{{2, 0}, {6, 1}}]]
]
```

Out[35]=

Let's introduce some style options to make the plot more colorful! Example 2.12.3 uses a different style option for each of the four objects. The argument to **Graphics** is now a list of four items, just as before, but now each item is itself a list consisting of the object together with a style attribute. Although this book is not printed in color, you will almost certainly want to use color in your graphics. Shades of gray can be gotten by using the **GrayLevel** function already described. Note that we have used a gray level of 0.5 to fill in the disk. To obtain true colors we use the **RGBColor**[*r, g, b*] function. Each of the parameters, r, g, and b are numbers between 0 and 1 that indicate how much of red, green, or blue to use. Thus using **RGBColor[1, 0, 0]**[4] in this example will produce a red polygon, which of course will look gray in this book. You need to try it yourself! For the circle we have introduced dashing and for the line a thickness.

Example 2.12.3

```
(* adding color and other options *)
Graphics[
 {{RGBColor[1, 0, 0],
   Polygon[{{0, 0}, {1, 2}, {3, 4}, {6, 2},
     {0, 0}}]},
  {Dashing[.02], Circle[{4, 5}, 1]},
  {GrayLevel[.5], Disk[{1, 4}, .5]},
  {Thickness[.02], Line[{{2, 0}, {6, 1}}]}}}
]
```

As a final example, we will create our very own Mondrian[5] painting by randomly drawing rectangles of random shapes and color in the plane. In Example 2.12.4, we have used the **Table** function to create a list of 400 rectangles that we then plot using

---

[4]Many colors can be given by simple names. **RGBColor[1, 0, 0]** is the same as **Red**. Look ahead to Chap. 6 for more on color names.

[5]Piet Mondrian (1872–1944) Dutch painter and member of the De Stijl art movement. See *Metamagical Themas* by Douglas R. Hofstadter, page 207, for an interesting comparison of actual Mondrian paintings to computer generated look-alikes.

the **Graphics** function. We have not used any options in the **Graphics** function. The trick was using the **RandomReal** function inside the **Rectangle** function. Let's start with the random function—**RandomReal**[x] will produce a random real number between 0 and x. The first thing we do is generate the lower-left corner of each rectangle randomly by using the point {**RandomReal[13], RandomReal[8]**}. This will give a point randomly chosen somewhere inside the rectangle of width 13 and height 8. The reason for choosing 13 and 8 is because 13/8 is pretty close to the Golden Ratio![6] Since we want to produce a work of art we may as well choose a well-proportioned canvas! After we choose the lower-left corner of each rectangle randomly we then want to choose the upper-right corner randomly too. But if we just chose a second point at random it might not lie to the right and above the first point that we chose! We get around this problem by letting x and y equal the coordinates of the first randomly chosen point. Then to produce the upper-right corner we add a random amount to both x and y. We don't want to make the rectangles too big as then they will overlap and fill up the whole canvas so we use **RandomReal[.4]** and **RandomReal[.5]**, respectively, to get the width and height of the rectangle. To summarize, the line

Rectangle[{x=RandomReal[13], y=RandomReal[8]},

{x+RandomReal[.4], y+RandomReal[.5]}]

creates the random rectangle and we then use the **Table** function to produce a list of 400 of these. But this is not all, since we want to also color the rectangles randomly too! So in fact, we are using the **Table** function to produce a list of pairs with each pair consisting of a randomly chosen color and a randomly chosen rectangle. Finally, to choose the color at random we use

RGBColor[RandomReal[ ], RandomReal[ ], RandomReal[ ]]

Here the three parameters r, g, b to the **RGBColor** function are being chosen at random by using **RandomReal[ ]**. We want each parameter to be a real number between 0 and 1, so we could have used **RandomReal[1]**. But if the argument to **RandomReal** is omitted *Mathematica* uses a value of 1.

---

[6]The Fibonnaci numbers are $0, 1, 1, 2, 3, 5, 8, 13, 21, \ldots$. Each number in the sequence is the sum of the previous two, and we start with zero and one. It turns out that the ratio of any number in the sequence to the previous number in the sequence converges to the Golden Ratio as we go further out in the sequence!

Example 2.12.4

```
In[36]:= (* creating a random drawing *)
Graphics[
 Table[
  {
   RGBColor[RandomReal[], RandomReal[],
    RandomReal[]], Rectangle[
    {x = RandomReal[13], y = RandomReal[8]},
    {x + RandomReal[.4], y + RandomReal[.5]}
   ]
  },
  {i, 1, 400}
 ]
]
```

Out[36]=

You should try making your own modern art by playing around with the number of rectangles (the 400) and the relative size of the canvas (the 13 and the 8) and the relative sizes of the rectangles (the .4 and the .5). But beware, each time you evaluate this cell you will get a new work of art, with your old masterpiece lost forever!

It turns out that the random number generator used by *Mathematica* does not really produce random numbers. In fact, writing a program to produce random numbers is not really possible. This topic is both mathematically and philosophically quite deep—way beyond the scope of this book! But, suffice it to say that each use of **RandomReal** will produce a sequence of numbers that depend only on the initial *seed*. The command **SeedRandom[*n*]** will reset the random generator using *n* as a seed. If we want to produce the same Mondrian painting each time we run Example 2.12.4, we could include **SeedRandom[1]**, for example, as the first line of code in that example. At the other extreme, **SeedRandom[ ]** will reset the generator in an almost unpredictable way, using the time of day and certain attributes from the current *Mathematica* session to derive the seed.

2.13 Saving and Printing Graphics

After putting a lot of effort into making a really cool graph of your favorite function complete with custom labels, coloring, dashing, and so on, you may want to print it, or save it so that you can paste into some other document or a Web site. First select the graphic by clicking on it or by clicking on its cell bracket. Next choose **File ▶ Print Selection** from the menu bar. Printing with *Mathematica* is similar to printing in other applications such as word processors. If you want to paste the graphic into another document or webpage try using **File ▶ Save Selection As** from the **File** menu. This will allow you to save the graphic in various formats such as PDF, HTML, JPEG, and the like. Or, after selecting the graphic you can also select **Edit ▶ Copy As** from the menu bar and then choose PDF or PICT. After copying the graphic you can then paste it into some other document. You'll have to experiment with your favorite applications to see what works best for you.

2.14 Find Out More

We have only discussed a few of the many graphics options that are available with the **Plot** command. There is a very nice tutorial, "Graphics and Sound," that you are urged to go through. Go to the Help Files (Documentation Center) and type "tutorial/GraphicsAndSoundOverview" in the search field. This should take you to the tutorial. Several of the **Plot** options that we did discuss also have additional optional features that we did not discuss. If you can imagine a way to jazz up your plot, there is probably an option that will do it! Just dig into the Help Files. The tutorial is a great place to start. Don't forget to evaluate **Options[Plot]** to see all the options available for **Plot**.

There is also a nice tutorial that will introduce you to the 2D Drawing palette and the Graphics Inspector. Go to the Help Files and type "tutorial/ InteractiveGraphicsOverview" into the search field.

We have only scratched the surface of what **Manipulate** can do! We'll be giving more examples in later chapters, but if you can't wait to see all the neat features it has, there is a nice tutorial in the Help Files. Search for "tutorial/Introduction-ToManipulate" and when you finish with that, check out "tutorial/Advanced-ManipulateFunctionality." Alternatively, type **?Manipulate** into a *Mathematica* notebook, and evaluate the cell. This will lead you into the Help Files.

It is also well worth taking a look at the Help File on **Graphics**. Go to the Help Files and search on "ref/Graphics" or enter **?Graphics**. The reference contains, among other things, the complete list of all graphics primitives and the correct syntax for their use.

Quiz

1. Use the **Plot** function to graph $-x^2 + x + 1$ from $x = -2$ to $x = 2$.

2. Repeat the first exercise, but use the **AspectRatio** option to make the plot have the same scale both vertically and horizontally.

3. Use the **Plot** function to simultaneously graph $-x^2 + x + 1$ and $x/2 - 1$, again from $x = -2$ to $x = 2$.

4. Repeat the last exercise, but make the two curves different thicknesses, or colors. Do this in two different ways:

   a. Use the Graphics Inspector. First open the the Graphics Inspector by selecting it from the **Graphics** menu. Next double-click on the graph of the parabola and modify it by using the controls in the Graphics Inspector.

   b. Don't use the Graphics Inspector. Instead, include a **PlotStyle** option in the **Plot** command.

   Which way do you find easier to use?

5. Continuing with the plot of the parabola and the line, use the Drawing Tools to add some labels to the plot.

6. Use **PolarPlot** to graph the polar equation $r = 1 + \cos\theta$ from $\theta = 0$ to $\theta = 2\pi$.

7. Modify the previous exercise by using **Manipulate** to graph $r = a + \cos\theta$ from $\theta = 0$ to $\theta = 2\pi$ and letting the parameter a range from $a = -2$ to $a = 2$.

8. Write a *Mathematica* cell that will draw the following:

   Hint: Use **Table** to create a list of six **Rectangles**.

9. Repeat the previous exercise but make each of the squares a different color.

10. Use **ListPlot** to plot 1000 randomly chosen points that all lie inside the square with vertices $(0, 0)$, $(1, 0)$, $(1, 1)$, and $(0, 1)$.

CHAPTER 3

Getting Help

In this chapter, we'll describe ways to avoid making mistakes with *Mathematica*. It makes sense to have this chapter early in the book so that you can, well, avoid making mistakes! On the other hand, in order to understand the examples, you'll need to know enough *Mathematica* to recognize the mistakes! So, try to read this chapter now—most of it can be read without yet knowing much *Mathematica*. And if any of it doesn't really make sense, just skip it. But, you should definitely come back and reread this chapter after you have learned more *Mathematica*. In fact, the more *Mathematica* you know, the more you will be able to appreciate this chapter and pick up some valuable tips for avoiding and correcting syntax errors as well as learning how to find answers to your *Mathematica* questions.

3.1 Syntax Coloring

You have probably already noticed that as you type expressions into an input cell, *Mathematica* will automatically color different parts of what you type different colors. This is being done to help you in a couple of important ways. First, it can help you identify, and avoid, syntax errors. Secondly, it can help you recognize and distinguish between variables, indices, and other expressions in ways that ultimately will help you organize your work.

Of course, this book is not printed in color! So as we look at the examples in this chapter it is doubly important to try them out yourself since the coloring we will be discussing will be rendered in shades of gray that may be difficult to see.

Let's look at a simple example. Suppose you want to create a list of squares and type the following into an input cell.

Example 3.1.1

```
(* syntax coloring helps find errors *)
Tabel[k^2, {k, 1, 10}]
```

Before you evaluate the cell it will look like it does above, with "Tabel" and both appearances of "k" in blue.[1] Notice also that the comment is colored gray. Having the comments a different color from the code makes them stand out and makes reading the code easier. Of course, the word "Tabel" is misspelled and is therefore unrecognized by *Mathematica* as a legitimate function. If we evaluate this cell nothing happens as we can see in the next example.

Example 3.1.2

```
In[6]:= (* since Tabel is undefined, nothing happens *)
        Tabel[k^2, {k, 1, 10}]

Out[6]= Tabel[k^2, {k, 1, 10}]
```

On the other hand, as we'll see in Chap. 5, there is nothing to keep you from creating your own function named **Tabel** in which case *Mathematica* would not have any trouble using it. It is worth noting that if you do define a function named **Tabel**, then all occurrences of "Tabel" in the notebook will immediately be recolored black. But, of course, you should not define a function named **Tabel** for at least three reasons. First, it's a good idea to use names that start with a lowercase letter for the functions you define since all *Mathematica* defined functions start with a capital letter. Secondly, "Tabel" is so obviously a misspelling of "Table" that it would be confusing to name a function "Tabel." Finally, names of functions should be as descriptive as possible to make reading the code (by you and by others) easier. Again, we'll discuss defining your own functions in Chap. 5.

---

[1] We'll see in Sec. 3.3 how you can change the color to whatever you want.

"Tabel" is obviously misspelled, but it may be a little less obvious what is wrong if you type **Arcsin[x]** instead of **ArcSin[x]**. So, syntax coloring can help you find errors and correct them.

Mathematica will automatically color blue any expressions that are not defined. So in Example 3.1.3, since *x* and *y* are symbolic variables that have no values, they remain colored blue.

Example 3.1.3

In[7]:= (* variables without values are colored blue *)
    Expand[(x + y)^5]

Out[7]= $x^5 + 5 x^4 y + 10 x^3 y^2 + 10 x^2 y^3 + 5 x y^4 + y^5$

If, after we enter the above cell, we give *x* a value by entering **x=2**, then the color of *x* in the above cell will turn to black. If we later clear the value of *x* by entering **Clear[x]**, all occurrences of *x* will return to blue.

There are a handful of syntax coloring conventions that *Mathematica* uses. So far we have seen that undefined expressions (functions, variables, and so on) are colored blue and comments are colored gray. Another very useful convention is that *Mathematica* will highlight missing delimiters. Consider Example 3.1.4, where we have accidentally left off the closing bracket.

Example 3.1.4

In[10]:= (* missing delimiters are highlighted *)   ⊞
    Tabel[k^2, {k, 1, 10}

Notice that the unmatched delimiter, in this case a left bracket, is highlighted in yellow, a plus sign in a yellow box is located at the right edge of the cell, and the cell bracket is highlighted in yellow. These highlightings are introduced when we try to evaluate the cell, and no output cell is produced. If we click on the plus sign in the yellow box, *Mathematica* beeps and we get a message that explains the problem. This is shown in Example 3.1.5.

Example 3.1.5

In[10]:= (* missing delimiter message *)   ⊞
    Tabel[k^2, {k, 1, 10}

    Syntax::bktmcp : Expression "Table[k^2, {k, 1, 10}" has no closing "]".
    Syntax::sntxi : Incomplete expression; more input is needed.

Mathematica will also make note of missing arguments. Suppose we want to make a list of the first 100 prime numbers and type the following into an input cell.

Example 3.1.6

```
(* Mathematica prompts for missing arguments *)
Table[Prime[i]∧]
```

The function **Prime[n]** will return the *n*th prime number. (This is a really cool function! Try entering **Prime[1000000000]**.) Of course, we have left off the indexing list {**k, 1, 100**} and *Mathematica* is prompting us for the missing argument by placing the caret symbol ∧ at the location of the missing argument.

In addition to showing us when arguments are missing, *Mathematica* will also let us know if we have entered too many arguments. In Example 3.1.7 the function **Length** should take one argument, but we have entered two. Look what happens.

Example 3.1.7

```
In[3]:= (* entering too many arguments leads to
    a warning *)
    Length[dataList, a]
```

       Length::argx : Length called with 2 arguments; 1 argument is expected. ≫

```
Out[3]= Length[dataList, a]
```

Here *Mathematica* has colored the extra argument and the comma that precedes it red as well as issued the descriptive warning.

We won't try to cover every syntax coloring convention, but as a final example, notice that *Mathematica* highlights arguments of user defined functions by coloring them gray and setting them in italics. In Example 3.1.8, we define our own function, named *f*, which squares its input and adds one. The argument, *x*, is colored gray and italicized.

Example 3.1.8

```
In[11]:= (* arguments of user defined functions
    are colored and italicized *)
    f[x_] := x^2 + 1
```

We will be discussing user defined functions in Chap. 5, so don't worry now if this example doesn't make total sense. The point is that *Mathematica* is highlighting the argument of the function to help us see what is going on.

3.2 Delimiter Matching

As you learn more about *Mathematica* and start to write more and more complicated code, it is not hard to have input cells that have dozens of lines. So far we haven't looked at any examples that are that complicated, but if you skip ahead in the book you will find a few pretty hairy-looking input cells! A very common error is to leave out matching delimiters. Even though *Mathematica* will try to warn us by highlighting missing delimiters as described in the previous section, it can sometimes still be hard to spot the error. A good habit to develop is to type the matching ending delimiter each time you type the beginning delimiter and then go back and fill in the missing information between the delimiters. This is especially helpful if the expression you are typing has many levels of nested delimiters. It is also pretty easy to do because *Mathematica* has keyboard shortcuts for inserting pairs of matching delimiters.

Let's look at a simple example. Suppose we want to type **Table[Prime[k], {k, 1, 100}]**. Of course, we could just type this from left to right and because it is not too long we'd probably be OK. But a good way to type it is as follows.

1. Type **Table**.
2. Select **Insert ▶ Typesetting ▶ Matching[ ]** from the menu bar. Actually, we want to use the keyboard shortcut for this menu item to enter the matching delimiters. If we actually had to stop typing and go to the menu with the mouse it wouldn't be convenient. We now have **Table[]**.
3. Use the left arrow key to move the cursor back so that it is between the bracket delimiters. Now type **Prime[k],** from left to right. (We could type **Prime**, insert the matching bracket delimiters, and back up to enter the **k**, but this is probably not worth it.)
4. Now use the keyboard shortcut to insert the matching delimiters {} (again available from the **Insert** menu). We now have **Table[Prime[k], {}]**. Use the left arrow key to back up and then type **k, 1, 100** to fill in the indexing list.

This may seem like a lot of trouble but it is actually not that many more key strokes then typing it from left to right, and doing it this way, we cannot forget to type closing delimiters that are needed to match opening delimiters. Typing this way is an excellent habit to develop and will pay off once you start to enter really big expressions. Again, for this relatively short example, it may not be worth it, but for longer examples it definitely is. And, if you get in the habit of typing this way, even short examples will fly off your fingertips.

Another nice feature for matching delimiters is that an opening delimiter will flash momentarily when the matching closing delimiter is typed. This feature is especially useful after you have been making changes to a long input cell with lots of nested delimiters, have messed it up terribly (often by trying to cut and paste code from somewhere else) and are searching for the opening delimiter that matches a given closing delimiter. Just erase the closing delimiter, retype it and look for the flashing opening delimiter! This is not the best way to go, but sometimes is just what you need.

A much better way to find matching delimiters, or look for a missing delimiter, is to use the menu item **Edit ▶ Check Balance**. If the cursor is placed anywhere in an input cell (or alternatively, if any portion of an input cell is selected), then **Check Balance** will expand the selection outward to cover the nearest pair of matched bracketing characters. Similar to **Check Balance** is **Extend Selection** which expands the selection outward to highlight the smallest subexpression containing the selection. **Extend Selection** can also be found in the **Edit** menu. Let's see how these can be used to find a missing delimiter.

Example 3.2.1 contains code with a missing delimiter. You don't need to understand what this code does now! We're just going to use it for this example. It is clear that there is a mismatched delimiter because *Mathematica* has highlighted the first opening bracket and has told us that the expression "has no closing]." But where do we need a closing bracket? (Not at the end!)

Example 3.2.1

```
In[35]:= nv = RandomInteger[{-10 000, 10 000}, 2];
    While[
     Length[
      pos =
       Position[tempTable =
         Table[nv.data[[i]], {i, 1, Length[data]}],
        Max[tempTable]] > 1,
      nv = RandomInteger[{-10 000, 10 000}, 2]
     ]
```

Syntax::bktmcp :
  Expression "While[Length [pos = Position[tempTable = «1», «1»] > 1, nv = «1»]" has no closing "]".

Syntax::bktmcp :
  Expression "While[Length [pos = Position[tempTable = «1», «1»] > 1, nv = «1»]" has no closing "]".

Syntax::sntxi : Incomplete expression; more input is needed.

Notice also that the expression **nv=RandomInteger[{-10000, 10000},2]** (on the eighth line) and the comma that precedes it is colored red. The last bracket in the cell is currently being paired with the opening bracket following **Length**, and *Mathematica* is coloring the subexpression red to indicate that we have too many arguments for the **Length** function.

Let's see how **Check Balance** or **Extend Selection** can help us find out where the missing delimiter belongs.

Once you learn a little more about *Mathematica* it will be easy to see that there is nothing wrong with the first line where we randomly define the two-dimensional vector **nv**. (Again, don't worry if this code makes no sense. We only want to use it to explain what **Extend Selection** does. We'll explain all the functions used in this example later in the book.) Also, the highlighted opening bracket is after the **While** function—more evidence that the first line is OK. If we place the cursor somewhere in the word **Table** and select **Extend Selection** we obtain the following:

```
While[
 Length[
  pos =
   Position[tempTable =
      Table[nv.data[[i]], {i, 1, Length[data]}],
     Max[tempTable]] > 1,
  nv = RandomInteger[{-10 000, 10 000}, 2]
 ]
```

Selecting **Extend Selection** again gives

```
While[
 Length[
  pos =
   Position[tempTable =
      Table[nv.data[[i]], {i, 1, Length[data]}],
     Max[tempTable]] > 1,
  nv = RandomInteger[{-10 000, 10 000}, 2]
 ]
```

At this point, with the entire table subexpression highlighted, we can see that there is nothing wrong with this subexpression—no missing delimiter yet. Applying **Extend Selection** a third time gives

```
While[
 Length[
  pos =
   Position[tempTable =
     Table[nv.data[[i]], {i, 1, Length[data]}],
    Max[tempTable]] > 1,
  nv = RandomInteger[{-10 000, 10 000}, 2]
 ]
```

OK, we are naming the table **tempTable** (for temporary table). Still no problem. Two more time gives:

```
While[
 Length[
  pos =
   Position[tempTable =
     Table[nv.data[[i]], {i, 1, Length[data]}],
    Max[tempTable]] > 1,
  nv = RandomInteger[{-10 000, 10 000}, 2]
 ]
```

At this point everything still seems fine. **Position** is a function that takes two arguments. The first is a list and the second is an element of the list. So here we are going to locate within the list **tempTable** its maximum element and the next thing we are going to do is name this location **pos**. If the maximum element appears in more than one position, then **Position** will list all the locations where it occurs and **pos** will have more than one element. Looking ahead, now we can see the problem! The **Length** function takes one argument, a list whose length it returns. The end of the **Length** function has to occur before the inequality sign. It doesn't make any sense to have an inequality sign as part of the argument to **Length**. The missing closing bracket goes right before the inequality sign. With it in that location, the **While** function now makes sense. (At least, after we understand what **While** does!) What this code does is choose the vector **nv** at random, use **Table** to form a list of the dot products of **nv** with each element of the list **data** and then see if this list of dot products has a unique maximum. If not, **nv** is chosen at random again and the process is repeated until the maximum dot product is unique.

Again, it really doesn't matter if you understand all the code in this example at this point. The goal is to see that **Extend Selection** can be used to repeatedly expand a subexpression while we are looking for (in this case) a missing delimiter. (You should try out **Check Balance** too. It works in a similar, but slightly different way.) As mentioned in the introduction to this chapter, the more you know about *Mathematica* the better able you will be to appreciate this chapter. This last example using **Extend Selection** falls into this category. So, if it didn't make total sense now, try rereading this part after you have more experience with *Mathematica*.

3.3 Setting Preferences

The color schemes that *Mathematica* uses for syntax highlighting as well as other aspects of the look and feel of your notebooks can be altered in two different ways. The first is to select **Mathematica ▶ Preferences** from the menu bar. There are quite a few parameters that are listed here that you are free to change. You should take a look and see if there is anything that sounds like it would be useful for you to change. We'll mention a few here.

Under "Appearance" and "Syntax Coloring" you'll see that you can change the colors that are used to mark local variables, comments, and the various subexpressions that are highlighted when *Mathematica* issues warnings. If you are red-green color blind like the author, you may want to switch these defaults to blues and yellows.[2]

Another way to change preferences, in ways that can apply to all of your *Mathematica* notebooks, or just the one you are presently working in, is to select **Format ▶ Option Inspector** from the menu bar. For example, suppose you want to change the "DelimiterFlashTime." Viewing the options by "Category," open "Editing Options," check "DelimiterFlashTime," and then select and change 0.3 seconds to say, 0.6 seconds. Another option you might want to change is the default magnification used in each notebook. Viewing the options "Alphabetically," for example, check "Magnification" and then set the value to say, 1.25. You can do this "Globally" so that all new notebooks will open with a magnification of 125%.

We encourage you to browse through the changes you can make with the **Option Inspector**. There may be several items that you would like to customize for your own use. We'll be coming back to the **Option Inspector** in Chap. 11 when we discuss changing the style of your cells or notebooks.

---

[2]Wouldn't it be cool if Wolfram incorporated two (or more) default coloring schemes into *Mathematica* so you could just open the "Preferences" and check "Red-Green Color blind" to get an appropriate selection of colors? According to the Howard Hughes Medical Institute (`http://www.hhmi.org/senses/b130.html`), about 7% of men in the United States, or about 10 million men, are red-green color blind. The rate is far less, about 0.4%, in women.

3.4 The Help Menu

Under the **Help** menu are several useful menu items. We have already mentioned the **Documentation Center, Function Navigator, Virtual Book**, and **Find Selected Function** in Chap. 1. In the **Documentation Center** you will find thousands of pages of documentation that we refer to simply as the "Help Files." The **Function Navigator** and **Virtual Book** lead into the Help Files. The Help Files are an invaluable tool that you simply must learn to navigate. We'll be pointing out useful guides and tutorials in the Help Files throughout the book.

We also mentioned in Chap. 1 that entering **?** followed by the name of a function, for example **?Plot**, will bring up a description of the syntax for that function, together with a hyperlink that will take you to the relevant page in the Help Files. This can also be accomplished by using the menu item **Help ▸ Find Selected Function**. For example, suppose **Plot** appears somewhere in our notebook. If we select the word **Plot** and then choose **Help ▸ Find Selected Function** (or better, use the keyboard shortcut) it will take us to the documentation page for **Plot**.

A variation on using **?Table** to find out about the **Table** command, is to use **?Ta***, where the asterisk acts like a wildcard that can stand for any text. Example 3.4.1 shows what happens if we enter **?Ta***.

Example 3.4.1

In[1]:= **? Ta***

▼ System`

Tab	TabSpacings	TagSetDelayed
TabFilling	TabView	TagStyle
Table	TabViewBox	TagUnset
TableAlignments	TabViewBoxOpti `. ons	Take
TableDepth	TagBox	TakeWhile
TableDirections	TagBoxNote	Tally
TableForm	TagBoxOptions	Tan
TableHeadings	TaggingRules	Tanh
TableSpacing	TagSet	TargetFunctions

We get a list of all functions that begin with the letters **Ta**. If we further click on one of the items in the list, it will bring up a brief description together with a hyperlink to the documentation for that function. At first it might seem that you already need to know what you are looking for in order to find it, but after using *Mathematica* for awhile there will be times when you almost remember something and this kind of wildcard searching is just what you need to quickly find it.

The position of the asterisk can be anywhere in the string, and can even be repeated. Try entering **?*String** or **?*String***.

Also in the **Help** menu are the items **Wolfram Website...** and **Demonstrations...** which, when selected, will take you to those Web sites. At the Wolfram site you can learn about other *Mathematica*-related products such as *gridMathematica, webMathematica* and *WolframWorkbench*, an integrated development environment that can be used to develop your own applications using *Mathematica*. Most of this stuff will not be of interest to *Mathematica* newbies, but you should definitely take a look at *The Mathematica Journal*, an online journal with lots of neat articles about *Mathematica*.

Much more interesting to the *Mathematica* novice (but useful for *Mathematica* experts too!), is the *Wolfram Demonstrations Project* which can be reached by selecting **Help ▶ Demonstrations....** Here you will find thousands of *Mathematica* notebooks, each of which *demonstrates* something interesting in mathematics or science. The demonstrations are written by *Mathematica* users who post them on the Web site so that they can be shared with others. After you learn how to use *Mathematica*, perhaps you will want to submit your own demonstration! This is a really great resource. By browsing the demonstrations you can often find one that either does exactly what you want to do, or something similar. You can download the code and modify it for your own purposes.

3.5 Writing Good Code

One of the absolute best things you can do to help yourself prevent errors, and find errors when you make them, is develop good habits of style. Using lots of comments in your code is a *very good idea*. Not only will it make your code easier to read by someone else,[3] you'll be amazed at how easy it is to forget what you were thinking yourself when you revisit some code you wrote even as little as a week or two ago.

---

[3]If you are a college or university student and using *Mathematica* to write up homework assignments, your Professors will be thrilled with your liberal use of comments!

We'll see in Chap. 11 that it is possible to enter text in *text cells* as opposed to the input cells that we have discussed so far. While you may want to use text cells to describe the calculations you are making in a notebook, they should be used differently than comments placed in input cells. For example, an advantage of placing comments inside an input cell versus a separate text cell, is that comments inside an input cell are not likely to get separated from the cell (if you cut and paste, for example) and hence lost.

Another extremely useful tool in writing easy to read code, and therefore more error-free code, is too adopt and consistently use an indentation convention. Consider the code we saw in Example 2.11.2 which will draw billiard trajectories. In Example 3.5.1 we give the same code, but without comments and without using any kind of indentation convention. Glance back to Example 2.11.2 and then look again at Example 3.5.1. Both input cells do the same thing, but clearly the first is much easier to read and hence to understand! The first example would also be much easier to troubleshoot if it happened to contain an error.

In Example 2.11.2, and in all the examples of this book, we try to follow (or *nearly* follow) the K&R indenting convention introduced by Kernighan and Ritchie in their classic book *The C Programming Language*. This convention is quite common now, not only in C programming, but in other languages too, and it can be used quite nicely with *Mathematica*. Of course, if we only have one line of code we don't need any indenting conventions: just type the code all on one line. But in Example 2.11.2 we have many lines of code. The basic convention is to type a function name, like **Manipulate**, and its opening delimiter on a single line and then indent the arguments to the function on subsequent lines. The final closing delimiter associated to the function is placed on a line by itself at the same indentation level as the function. In Example 2.11.2 only **Manipulate[** and its final closing **]** are at the leftmost indentation level. Between these lines are the arguments to **Manipulate**. The first argument is **ParametricPlot** for which the convention is repeated. We see **ParametricPlot[** on one line, then all of its arguments indented on subsequent lines, and then the final closing bracket to **ParametricPlot** in a line by itself at the same indentation level as **ParametricPlot**. Within **ParametricPlot** some of the arguments, like the first one, are too long to fit on one line, so it is split across two at a natural splitting point. The first three options, **PlotRange, Axes**, and **PerformanceGoal** are each placed on a separate line. The **Epilog** option is too long to fit on a single line and is split in an organized way. Notice that **Line**, which appears as an element of the **Epilog** list, is treated with the K&R style: **Line[** is placed on one line, then its argument on an indented line, and finally the closing bracket to **Line** on a line by itself at the same indentation level as **Line**.

Compare reading this with Example 3.5.1! The examples clearly speak for themselves.

Example 3.5.1

```
(* badly written code! *)
Manipulate[ParametricPlot[{1 - Abs[Mod[
        hspeed t + pt[[1]], 2] - 1],
   1 - Abs[Mod[vspeed t + pt[[2]], 2] - 1]}, {t, 0,
   pathLength}, PlotRange → {{-.01, 1.01}, {-.01, 1.01}},
  Axes →
   False, PerformanceGoal → "Quality", Epilog →
   {Thickness[.01], Line[{{0, 0}, {0, 1}, {1, 1},
       {1, 0}, {0, 0}}]}]], Style["Speed", 12, Bold],
 {{hspeed, .2, "Horizontal"}, -1, 1}, {{vspeed,
   .5, "Vertical"}, -1, 1}, Delimiter, Style["Initial
Position", 12 , Bold], {{pt, {.5, .5}, ""}, {0, 0},
  {1, 1}}, Delimiter, Style["Length", 12 , Bold],
 {{pathLength, 5, ""}, .01, 100}]
```

Mathematica will not automatically force you to type your code with the K&R style, but *Mathematica* does provide automatic indenting when a subexpression is spread over more than one line. So it is easy to use K&R style with *Mathematica*.

Of course, the disadvantage of using K&R style is that more lines are needed, although this is not really much of a disadvantage. Sometimes it is possible to break with the convention slightly without losing the effectiveness of the style. For example, we might place **p=3;q=5** on a single line rather than **p=3** and **q=5** on separate lines without losing any readability of the code. In the end you will develop your own style as you use *Mathematica* more and more. Looking through the examples, you'll notice that the author's style does not always follow K&R style exactly. But the point is that developing and using a consistent style is a hallmark of good programming and can be a source of pride.

In addition to the indenting convention, notice the names of the variables we used in Example 2.11.2: **hspeed** and **vspeed** for the horizontal and vertical speed; **t** for time; **pt** for the initial point; **pathLength** for the length of the path. You should choose variable names that are descriptive. Perhaps in this case we should have used **initialPoint** instead of **pt**. Notice also the use of the upper case **L** in **pathLength**. This makes the string easier to parse as two words. Perhaps we should have used **hSpeed** and **vSpeed**. Finally, none of the variables we defined start with uppercase letters. We leave that convention for *Mathematica* to use with its built-in functions.

3.6 Monitoring Calculations

We have already mentioned the **Timing** function in Chap. 1. We'll be doing some very time consuming calculations in Chap. 10 where **Timing** can be quite useful.

A more dynamic way to see how long a calculation takes is to use **ProgressIndicator** with the **Monitor** function. Suppose we want to factor all the integers from 1 to 10 000 000 into their prime factors. Obviously, this might take some time! Example 3.6.1 will do just this and place the results in a list. Let's see how it works and how we can use **ProgressIndicator** to monitor how the calculation is going. Consider the following Example 3.6.1.[4]

Example 3.6.1

```
In[65]:= (* monitoring the progress of a calculation *)
    Monitor[
     Table[
       FactorInteger[i],
       {i, 1, 10 000 000}
      ];,
     ProgressIndicator[i, {1, 10 000 000}]
    ]
```

The effect of **ProgressIndicator** is to display a "speedometer bar," or progress indicator, that indicates how far along the calculation is while it is in progress. After the calculation is complete, the progress indicator is no longer displayed. This way we can see how close we are to being done.

Let's start with the **Table** function. We are using this to make the list of factored integers. The indexing list for **Table** is {**i, 1, 10 000 000**} so that our list will have 10 000 000 elements. Furthermore, the *i*th element will be **FactorInteger[i]**, which we'll explain further in a moment. Note that we have placed a semicolon after the **Table** function to suppress the output of the list. We really don't want to print out a list with 10 000 000 entries!

To understand what **FactorInteger** does, let's just factor the first 10 positive integers. We do this in Example 3.6.2. Notice that we have removed the semicolon from after the **Table** function. We'll talk about **Grid** in just a moment. The output of

---

[4]Warning: If you try to run this example it will take a very long time! Try running it first with the 10 000 000 lowered to 10 000. The point is to have a calculation that takes long enough so that you can watch the progress indicator.

Example 3.6.2 looks a little strange but makes sense once we know that **FactorInteger[i]** returns a list of all prime factors of *i* together with their multiplicities. Thus 8, which is 2^3, factors as {{2, 3}}, and 6, which is $2 \cdot 3$, factors as {{2, 1}, {3, 1}}. If we factored $2^3 \cdot 3^5 \cdot 17$ we would get {{2, 3}, {3, 5}, {17, 1}}.

Example 3.6.2

```
In[22]:= (* factoring the first 10 positive integers *)
    Grid[
     Table[
      FactorInteger[i],
       {i, 1, 10}
      ]
     ]
```

```
        {1, 1}
        {2, 1}
        {3, 1}
        {2, 2}
        {5, 1}
Out[22]=  {2, 1}   {3, 1}
        {7, 1}
        {2, 3}
        {3, 2}
        {2, 1}   {5, 1}
```

We don't need to use **Grid** here, but it formats the output nicely. In general, **Grid** takes a list of lists and prints the elements of each element of the big list on each line. The last sentence was correct, but probably hard to understand! Try running this example without **Grid**, just using the **Table** function. Now you will see the list of lists that is produced. Compare this with the output using **Grid** and you will see that each line of output in Example 3.6.2 contains the elements of a list that was itself a single element of the big list. It's hard to say, but easy to see once you understand it!

Returning to Example 3.6.1, what we have done is wrap the **Monitor** function around the **Table** function, and we have used **ProgressIndicator** to produce the progress indicator that we want. The function **ProgressIndicator** can be used to produce a "bar graph" (in the form of a shaded speedometer bar) of any numerical expression, with a list of the minimum and maximum values given as the second argument. The **Monitor** function takes two arguments. The first is the calculation that we want to monitor, and the second is what we want to display to monitor the calculation. Instead of using the progress indicator as we did in this example, try letting the second argument to **Monitor** simply be **i**. In this case the integer *i* will be displayed as the calculation takes place. The effect is like looking at a car's odometer quickly turning over. Try it!

If we slow a calculation down it can be easier to see whatever it is that we are monitoring. One of the Quiz questions shows you how to do this with the **Pause** function.

3.7 Getting Out of Trouble

Sometimes you'll do something that seems to really mess up everything. For example, you might evaluate a cell and *Mathematica* just keeps running and running and clearly something is amiss. As we have already mentioned in Chap. 1, try selecting **Evaluation ▶ Abort Calculation** from the menu bar to stop a calculation. Unfortunately, there are times when this will not work. (Sometimes it will work if you just keep trying it over and over.) If **Abort Calculation** will not work, you can try **Evaluation ▶ Quit Kernal/Local**. The *Mathematica* Kernal is the mathematical engine that runs behind the scene, carrying out all of your calculations. Everything that you see: the notebook, input and output cells, and so on, is the *Front End*. Choosing **Quit Kernal** will stop the kernal dead in its tracks without stopping the Front End and closing your Notebook. This is a better alternative to just quitting *Mathematica*, in which case you will lose any changes made to your notebook since your last **Save**. A good habit to develop with *Mathematica*, as with most computer applications, is to **Save** your work regularly.

If you do quit the kernal, it will restart as soon as you evaluate a cell in your notebook.

Another thing that can go wrong is that you may make a mistake which causes some variables or functions to be defined in ways that you are not really aware of. If a function or variable just does not seem to be behaving properly, try using the **Clear** function to *undefine* it and then start over. This can often get you out of a jam.

3.8 Ask an Expert

Of course, one of the best ways to learn about anything is to learn from an expert. If you are having a problem with *Mathematica* that you just can't figure out, and you have already spent time searching the Help Files, it's great to find someone who knows exactly what to do. If you are a student or professor at a college or university, see if anyone on the Information Technology staff, or computer lab staff can help you. Or perhaps your community has a local *Mathematica* Users Group.

The internet is also a great resource. If you *Google* your *Mathematica* question you just might find someone who has posted the answer on a Web site. Finally, as you peruse the Wolfram Web site you'll see that they offer courses in *Mathematica*. Some are free, online, real-time lectures while others are several-day workshops.

3.9 Find Out More

A couple of guides and tutorials in the Help Files worth taking a look at are

- tutorial/WarningsAndMessages
- tutorial/OptionInspector

Mathematica also includes a debugger that can be found under the **Evaluation** menu. As you become more and more expert at *Mathematica* and start writing more complicated code, the debugger can be a real life saver. Since most beginners won't need it, we won't talk about it in this book.

Quiz

1. Find the missing delimiter in the following code:

```
Plot[
    E^x,
    {x, -1, 10},
    PlotRange → {{-1, 4}, {0, 10},
        PlotLabel → "The Graph of e^x."
        ]
```

2. Change the default color for comments in all your notebooks to be blue.

3. Rewrite the following code with the K&R indenting scheme. This code will find all prime numbers from 2 to *max* by using the famous Sieve of Eratosthenes algorithm. The code uses a *Do loop* which we will be discussing in Chap. 5, but since the question here is simply to format the code, you don't really need to know what a Do loop is yet! (If you prefer to wait and answer this question, and the next one, after you read Chap. 5, that's fine.) (* **Sieve of Eratosthenes** *) (* **initialize the sieve: make a list of the integers from 1 to max** *) **max = 100; sieve = Table[i, {i, 1, max}]; (* now for each integer k from 2 to max/2, see if it has been crossed out. If not, cross out all of its multiples. A number in the list will be "crossed out" if it has been**

replaced with a zero *) Do[(* see if the number k has been crossed out *) If[sieve[[k]] != 0, (* cross out all its multiples *) j = 2; While[j k <= max, sieve[[j k]] = 0; j++]] , {k, 2, max/2}] (* the sieve now contains the primes and zeroes. Union will remove duplicate zeroes and sort the list. Drop removes the 0 and 1 that are now the first two elements *) primes = Drop[Union[sieve], 2]

4. Get the Sieve of Eratosthenes code in the previous example running and use **Monitor** with a **ProgressIndicator** to monitor its progress as it runs. If you change **max** to be 1 000 000 the runtime should still be modest.

5. Change Example 3.6.1 as follows: Replace **FactorInteger[i]** with **Pause[1]**, replace {**i, 1, 10 000 000**} with {**i,1,10**}, and replace **ProgressIndicator[i,{1,10000000}]** with {**i, FactorInteger[i]**}. What happens? Look up the **Pause** function in the Help Files to see exactly what it does. What happens if you remove the semicolon after the **Table** function?

CHAPTER 4

Odds and Ends

Unlike the other chapters in this book, which are all focused on a single topic, this chapter will introduce a mix of functions that are useful in all sorts of situations.

4.1 Transforming Expressions

Quite often it is useful to put a mathematical expression into a different form. Perhaps looking at the expression in a different way will lend some critical insight. *Mathematica* has several functions that can help us do this. In Example 4.1.1, we use the **Expand** function to multiply out a product and name the result **poly**.

Example 4.1.1

In[15]:= (* expanding a polynomial *)

    poly = Expand$\left[(1 + x)^3 \ (x - 2 \ y) \right]$

Out[15]= $x + 3 \ x^2 + 3 \ x^3 + x^4 - 2 \ y - 6 \ x \ y - 6 \ x^2 \ y - 2 \ x^3 \ y$

We can perform the opposite of **Expand** by using the **Factor** function. In Example 4.1.2, we factor **poly** and get right back to where we started.

Example 4.1.2

In[16]:= (* factoring a polynomial *)
        Factor[poly]

Out[16]= $(1 + x)^3 (x - 2 y)$

Since **poly** is a polynomial in both x and y, it might be interesting to write it as a polynomial in one variable, with coefficients in the other. We can do this with the **Collect** function. In general, **Collect** takes two arguments. The first is the expression that we want to transform and the second is the variable with respect to which we want to collect terms.

Example 4.1.3

In[17]:= (* poly is a polynomial in y with
           coefficients in x *)
        Collect[poly, y]
        Collect[poly, y, Simplify]

Out[17]= $x + 3 x^2 + 3 x^3 + x^4 + \left(-2 - 6 x - 6 x^2 - 2 x^3\right) y$

Out[18]= $x (1 + x)^3 - 2 (1 + x)^3 y$

In Example 4.1.3, we **Collect** the terms of **poly** with respect to **y**. The polynomial is linear with respect to y, but the coefficients are themselves polynomials in x. In the second usage of **Collect** above, we add the option **Simplify** which then simplifies each of the coefficient polynomials, in this case factoring them.

In fact, **Simplify** is a powerful function its own right and using **Simplify** will cause *Mathematica* to return an expression it is "simplest" form. Unfortunately, there is no hard and fast rule as to when an expression is in "simplest" form, so what we get may or may not be the most useful form for whatever it is that we are trying to do. In the case of **poly**, the next example shows that **Simplify** considers the factored form to be the simplest representation of the expression.

Example 4.1.4

In[19]:= (* simplifying the polynomial *)
        Simplify[poly]

Out[19]= $(1 + x)^3 (x - 2 y)$

Notice that **poly** is divisible by $1 + x$. Let's take a look at the quotient. In Example 4.1.5, we try to expand the polynomial divided by $1 + x$ with some rather unanticipated results.

Example 4.1.5

In[20]:= (* poly is divisible by 1+x,
     but here the numerator is expanded! *)
   Expand[poly / (1 + x)]

Out[20]= $\dfrac{x}{1+x} + \dfrac{3\,x^2}{1+x} + \dfrac{3\,x^3}{1+x} + \dfrac{x^4}{1+x} -$

$\dfrac{2\,y}{1+x} - \dfrac{6\,x\,y}{1+x} - \dfrac{6\,x^2\,y}{1+x} - \dfrac{2\,x^3\,y}{1+x}$

Expand did not perform the division and expand the quotient! Instead, as it does with all rational expressions,[1] it expanded the numerator and left the denominator alone. Example 4.1.6 shows how we can actually do the division and then expand what is left. Both **Factor** and **Simplify** will do the same thing in this case, namely, cancel the common factor of $1 + x$. After performing the cancellation we can then use **Expand** to multiply out the quotient.

Example 4.1.6

In[21]:= (* expanding the quotient poly/(1+x) *)
   Factor[poly / (1 + x)]
   Expand[Factor[poly / (1 + x)]]

Out[21]= $(1 + x)^2\,(x - 2\,y)$

Out[22]= $x + 2\,x^2 + x^3 - 2\,y - 4\,x\,y - 2\,x^2\,y$

Another way to do this is to use the **Cancel** function, which cancels common factors in quotients. Notice that **Cancel** returns the expanded quotient, not the factored quotient, in this case.

Example 4.1.7

In[23]:= (* using Cancel to express the quotient *)
   Cancel[poly / (1 + x)]

Out[23]= $x + 2\,x^2 + x^3 - 2\,y - 4\,x\,y - 2\,x^2\,y$

---

[1] A *rational* expression is a quotient of polynomials.

Simplify is aware of all kinds of useful identities that might be used to simplify an expression. For example, in trigonometry, you probably remember the *Pythagorean Identity*, $\cos^2 x + \sin^2 x = 1$, and also remember that there are a whole lot of other *trigonometric identities*. These identities can often be used to simplify incredibly complicated trigonometric expressions. (Conversely, they can be used to horribly mess up simple expressions!) Look at what a great job **Simplify** does with the following trigonometric expressions.

Example 4.1.8

In[1]:= (* Simplify can use trig identities *)
Simplify$\left[5\, \text{Cos}[\Theta]^4\, \text{Sin}[\Theta] - 10\, \text{Cos}[\Theta]^2\, \text{Sin}[\Theta]^3 + \text{Sin}[\Theta]^5\right]$

Simplify$\left[\dfrac{3 - 4\,\text{Cos}[2\,\Theta] + \text{Cos}[4\,\Theta]}{4\,(1 + \text{Cos}[2\,\Theta])}\right]$

Out[1]= $\text{Sin}[5\,\Theta]$

Out[2]= $\text{Sin}[\Theta]^2\, \text{Tan}[\Theta]^2$

Playing around with **Factor, Expand, Simplify**, and **Cancel** will often change expressions in ways that are helpful. On the Algebraic Manipulation Palette, pictured in Fig. 4.1, you'll find quite a few more functions that can be used to manipulate expressions. When working with expressions that involve trigonometric functions you will find **TrigExpand, TrigFactor**, and **TrigReduce** (in addition to **Simplify**) especially useful. You can access the palette from the **Palettes** menu, and click on a palette item to save typing it yourself.

Let's look at a few more of these functions. Suppose we want to simplify

$$\frac{x^3 y^{\frac{4}{3}}}{z w^5} \sqrt{\frac{x^3 w^5}{y}}$$

Example 4.1.9 illustrates what **Simplify** as well as **FullSimplify** do to this expression. Hmmm... What they give doesn't look all that much better than what we started with. Generally speaking, **FullSimplify** will do a better job than **Simplify** (at the expense of being slower), so if **Simplify** doesn't seem to do what you want, try **FullSimplify**. But, in this case, it didn't help.

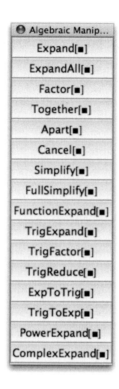

Figure 4.1   The Algebraic Manipulation palette contains many functions that can be used to transform expressions.

Example 4.1.9

In[266]:= (* trying to simplify an expression *)

$$\text{Simplify}\left[\frac{x^3\,y^{\frac{4}{3}}}{z\,w^5}\,\sqrt{\frac{x^3\,w^5}{y}}\,\right]$$

$$\text{FullSimplify}\left[\frac{x^3\,y^{\frac{4}{3}}}{z\,w^5}\,\sqrt{\frac{x^3\,w^5}{y}}\,\right]$$

Out[266]= $\dfrac{x^6\,y^{1/3}}{\sqrt{\frac{w^5\,x^3}{y}}\;z}$

Out[267]= $\dfrac{x^6\,y^{1/3}}{\sqrt{\frac{w^5\,x^3}{y}}\;z}$

Instead, let's try **PowerExpand**. What this function does is expand all powers of products using the rules of exponents. Perhaps it will be able to combine all the powers of, for example x, into a single power. In addition to tackling the expression of Example 4.1.9, we give a couple of other examples below. Well, this looks pretty good!

Example 4.1.10

```
In[271]:= (* PowerExpand will use rules of exponents to
          simplify expressions *)
      PowerExpand[(a b)^c]
      PowerExpand[(a^b)^c]
```

$$\text{PowerExpand}\left[\frac{x^3\, y^{\frac{4}{3}}}{z\, w^5}\sqrt{\frac{x^3\, w^5}{y}}\right]$$

Out[271]= $a^c\, b^c$

Out[272]= $a^{b\,c}$

Out[273]= $\dfrac{x^{9/2}\, y^{5/6}}{w^{5/2}\, z}$

But we need to be careful! In Example 4.1.11, we enter **PowerExpand**[$\sqrt{x^2}$] and get x. At first this seems like it is correct, and seems much better than what we get if we enter **Simplify**[$\sqrt{x}$], which doesn't do anything. But, unfortunately, the output of **PowerExpand** in this case is not quite correct.

Example 4.1.11

```
      (* this is not completely correct *)
```

$$\text{PowerExpand}\left[\sqrt{x^2}\right]$$

Out[274]= x

```
In[275]:= (* this is correct *)
```

$$\text{Simplify}\left[\sqrt{x^2}\right]$$

Out[275]= $\sqrt{x^2}$

If x is a real number, $\sqrt{x^2}$ is only x if $x \geq 0$. Otherwise it is $-x$. For example, if $x = -3$ then $x^2 = 9$ and $\sqrt{x^2} = \sqrt{9} = 3$ which is not x. In general, $\sqrt{x^2} = |x|$, the absolute value of x.[2] (And, if x is a complex number, then even this is not true.) Luckily, we can let *Mathematica* know if the variables we are working with represent positive or negative numbers, (or even real numbers) in which case both **PowerExpand** and **Simplify** can do a better job. In the next Example, we repeat Example 4.1.11, but use the option **Assumptions** to indicate the domain of the variables.

Example 4.1.12

In[276]:= (* using Assumptions to restrict the domain
        of the variable *)

PowerExpand$\left[\sqrt{x^2}, \text{Assumptions} \rightarrow \{x \in \text{Reals}\}\right]$

Simplify$\left[\sqrt{x^2}, \text{Assumptions} \rightarrow \{x \in \text{Reals}\}\right]$

Simplify$\left[\sqrt{x^2}, \text{Assumptions} \rightarrow \{x < 0\}\right]$

Out[276]= $e^{i\pi \text{Floor}\left[\frac{1}{2} - \frac{\text{Arg}[x]}{\pi}\right]} x$

Out[277]= Abs$[x]$

Out[278]= $-x$

In the first two cases, we use the **Assumptions** option to tell *Mathematica* that x is a real number. The symbol $\in$ means "is an element of" and so $x \in$ **Reals** means that x is a member of the set of real numbers. You can find the $\in$ symbol in the BasicMathInput palette. In the last example, we use the **Assumptions** option to declare that x is negative (and hence a real number). Given these assumptions, the functions now return correct answers.

However, the first answer is a bit cryptic. Given any complex number z, its *argument*, or arg z, is the angle through which we must rotate the plane around the origin to bring z onto the positive real axis. Every complex number z can be written as $z = re^{i\theta} = r(\cos\theta + i\sin\theta)$ where $r = |z|$ is the absolute value, or *norm*, of z and $\theta = \arg z$. A positive real number has an argument of zero, while a negative real number has an argument of π radians. Thus if $x > 0$, **Floor**$[\frac{1}{2} - \frac{\text{Arg}[x]}{\pi}]$ will be **Floor**$[\frac{1}{2}]$ which is zero. The **Floor** of a real number x is the largest integer less than

---

[2]Remember that $\sqrt{x}$ stands for the positive number whose square is x. If we want to represent the negative number whose square is x, we use $-\sqrt{x}$.

or equal to x. (Related to the **Floor** function is the **Ceiling** function—**Ceiling**[x] is the smallest integer greater than or equal to x.) Finally, if $x > 0$, we get $e^0 x = x$. On the other hand, if $x < 0$, Arg[x] $= \pi$ and we end up with $e^{-i\pi} x = -x$. So, in fact, **PowerExpand** has returned a correct answer, although not as simple as what **Simplify** returns.

Trying to manipulate and simplify expressions with the *Mathematica* functions given in the AlgebraicManipulation palette is a bit of an art form. You simply need to start experimenting with these functions until you get the hang of how they work. Sometimes one of them will do just what you need while others do nothing or make matters worse. Just keep trying, and keep looking in the Help Files for examples and advice.

4.2 Replacement Rules

Sometimes we need to substitute something for a variable within a given algebraic expression. We can do this with *Mathematica* by using *replacement rules*. For example, suppose we are working with the 2-variable polynomial given in the first section and we want to replace x with 2. One way to do this is to enter **x = 2**. Now whenever x is encountered it will be replaced with 2. If we then re-enter **poly** we'll get an expression in y alone obtained by setting x equal to 2. This may be just what we want, but on the other hand, perhaps we don't want to permanently set x to be 2.

An alternative, one that will not permanently assign the value 2 to x, is shown below. Here we have used the **ReplaceAll** function by following the expression with the *slash-dot*, **/.**, after which we give the *replacement rule*.

Example 4.2.1

In[27]:= (* replacing x with 2 in poly *)
    poly /. x → 2

Out[27]= 54 − 54 y

The real power in the slash-dot replacement function lies in the fact that we can replace one thing with any other thing. In general, if we evaluate *expression /. lhs → rhs*, *Mathematica* will make one pass through the expression replacing each occurrence of *lhs* with *rhs*.[3]

---
[3]The strings *lhs* and *rhs* stand for "left-hand side" and "right-hand side," respectively.

We can also replace more than one thing at the same time. In the following example we replace both x and y in **poly**. We replace x with $a + 1$ and y with $b - 2$. Note that the replacement rule has become a list of replacement rules, one for each expression that we wish to replace.

Example 4.2.2

```
In[29]:= (* replacing both x and y with other
          expressions *)
       poly /. {x → a + 1, y → b - 2}
```
Out[29]= $1 + a + 3 (1 + a)^2 + 3 (1 + a)^3 +$
         $(1 + a)^4 - 2 (-2 + b) - 6 (1 + a) (-2 + b) -$
         $6 (1 + a)^2 (-2 + b) - 2 (1 + a)^3 (-2 + b)$

After the replacement is made, the expression is not simplified in any way, so in Example 4.2.2 we might want to follow the replacement with **Expand** or **Simplify** or **Factor**.

Let's use this to do a real problem. Suppose we want to prove that the polynomial

$$-829 + 1575y - 1245y^2 + 525y^3 - 120y^4 + 12y^5$$

cannot be factored.[4] Of course, we can try to factor it with *Mathematica*. In Example 4.2.3, we name the polynomial **A** and then use **Factor[A]** only to find that it does not factor.

Example 4.2.3

```
In[2]:= (* this polynomial cannot be factored *)
       A = -829 + 1575 y - 1245 y² + 525 y³ - 120 y⁴ + 12 y⁵;
       Factor[A]
```
Out[3]= $-829 + 1575 y - 1245 y^2 + 525 y^3 - 120 y^4 + 12 y^5$

At this point we should truly believe the polynomial is *irreducible*, that is, cannot be factored, since *Mathematica*'s algorithm for factoring polynomials should be foolproof on only a fifth degree polynomial! But suppose we need to prove it is irreducible.[5] There is a nice test, known as Eisenstein's criterion, that can sometimes

---

[4] Only a mathematician would think this was a "real" problem!

[5] Again, who but a mathematician would feel the need!

be used to show that a polynomial is irreducible. It says that if there is a prime number p that divides all the coefficients of the polynomial except that of the highest degree term, and furthermore p^2 does not divide the constant term, then the polynomial is irreducible. Let's try this with **A**. It would help a lot if we factored the coefficients of **A**, which we do in the next example.

Example 4.2.4

In[15]:=

```
(* factoring the coefficients of A *)
cList = CoefficientList[A, y]
FactorInteger[cList]
```

Out[15]= {-829, 1575, -1245, 525, -120, 12}

Out[16]= {{{-1, 1}, {829, 1}}, {{3, 2}, {5, 2}, {7, 1}},
        {{-1, 1}, {3, 1}, {5, 1}, {83, 1}},
        {{3, 1}, {5, 2}, {7, 1}},
        {{-1, 1}, {2, 3}, {3, 1}, {5, 1}},
        {{2, 2}, {3, 1}}}
```

---

In Example 4.2.4 we have used **CoefficientList[A, y]** to form a list of the coefficients which we then name **cList**. We enter two arguments for **CoefficientList**: the polynomial and then the variable of the polynomial. We have already seen **FactorInteger** in Chap. 3 in Examples 3.6.1 and 3.6.2. But notice that here we are applying **FactorInteger** not to a single integer, but to a whole list of integers. In this case the function is applied individually to each number in the list. We say that **FactorInteger** is a *listable* function. Most functions in *Mathematica* are listable and we'll be saying more about this property later in this chapter.

Returning to Example 4.2.4, we see that the first coefficient, $-829$, is prime. So Eisenstein's criterion does not work. The only prime that we could possibly use is 829, but this prime does not divide any of the other coefficients.

But, all hope is not lost! If we replace $y$ with $x + k$, where $k$ is some integer we will get a new polynomial with new coefficients, but the new polynomial will factor if and only if the original one factors.[6] So maybe we can do such a substitution and obtain a polynomial for which Eisenstein's criterion works! To employ this strategy, let's replace $y$ with $x + k$ for various values of $k$ and see what we get. In Example 4.2.5, we replace $y$ with $x + 1$ and factor the coefficients again. The first coefficient is only divisible by the primes 2 and 41. But, the

---

[6]Think about it!
```

fourth coefficient, 165, is not divisible by either of these. So, again, the method fails.

Example 4.2.5

In[21]:= (* replacing y with x+1 *)
        B = Expand[A /. y → x + 1]
        FactorInteger[CoefficientList[B, x]]

Out[21]= $-82 + 240\, x - 270\, x^2 + 165\, x^3 - 60\, x^4 + 12\, x^5$

Out[22]= {{{-1, 1}, {2, 1}, {41, 1}},
        {{2, 4}, {3, 1}, {5, 1}},
        {{-1, 1}, {2, 1}, {3, 3}, {5, 1}},
        {{3, 1}, {5, 1}, {11, 1}},
        {{-1, 1}, {2, 2}, {3, 1}, {5, 1}},
        {{2, 2}, {3, 1}}}

Let's try one more time, but this time, systematically test different values of k. In Example 4.2.6, we use **Table** to make a list of polynomials each one obtained from **A** by replacing y with $x + k$, where k runs from -2 to 2. After doing the replacement we use **Expand** to multiply out the expression. Finally, we use the function **TableForm** to print one polynomial on each line. In general, **TableForm** takes a list as its argument and then prints the elements in a nice array.

Example 4.2.6

In[27]:= (* trying different linear substitutions *)
        TableForm[
        Table[
          Expand[A /. y → x + k],
          {k, -2, 2}
        ]
        ]

Out[27]//TableForm=

$-15\,463 + 17\,655\, x - 8235\, x^2 + 1965\, x^3 - 240\, x^4 + 12\, x^5$

$-4306 + 6180\, x - 3660\, x^2 + 1125\, x^3 - 180\, x^4 + 12\, x^5$

$-829 + 1575\, x - 1245\, x^2 + 525\, x^3 - 120\, x^4 + 12\, x^5$

$-82 + 240\, x - 270\, x^2 + 165\, x^3 - 60\, x^4 + 12\, x^5$

$5 + 15\, x - 15\, x^2 + 45\, x^3 + 12\, x^5$

It's pretty hard to tell if the first two polynomials (where $y = x - 2$ and $y = x - 1$) satisfy Eisenstein's criterion. We'd have to factor the coefficients to find out. But looking ahead to the last polynomial, the one with $y = x + 2$, we can see that the prime 5 works! It divides every coefficient except 12, and its square, 25, does not divide the constant term. Hooray! We have shown that the original polynomial **A** is irreducible. (And so are all the others in the list.)

The slash-dot replacement function is incredibly useful so it is important to understand exactly how it works. When we enter *expression /. rules*, where *rules* might be a list of replacement rules, each rule is applied to each part of the expression until a rule is found that matches. That substitution is made and then the rules are applied again, in order, to the next part of the expression. So what will happen to **a/.{a→2, a→3}**? The rules are contradictory. Should we replace **a** with 2 or with 3? Since the rules are applied in order, **a** will be replaced by 2 and the rule **a→3** is never used.

Furthermore, only one pass through the expression is made. Suppose we enter **x+y → {x→y, y→2}**. The **x** will be replaced by **y** and the **y** will be replaced with 2 and we will get **y+2**. This is not the same as **x+y/.x→y/.y→2** which will yield 4. Example 4.2.7 illustrates this.

Example 4.2.7

In[44]:= (* these substitutions are not the same *)
        x + y /. {x → y, y → 2}
        x + y /. x → y /. y → 2

Out[44]= 2 + y

Out[45]= 4

Since only one pass is made through the expression, this provides a very handy way for swapping variables. In Example 4.2.8 we start with an expression in x and y and replace it with one where the variables have been traded.

Example 4.2.8

In[47]:= (* swapping x and y *)

$$\frac{x \, \text{Sin}[x \, y]}{\sqrt{x^2 + y^3}} \; /. \; \{x → y, \; y → x\}$$

Out[47]= $\dfrac{y \, \text{Sin}[x \, y]}{\sqrt{x^3 + y^2}}$

Finally, there may be times when you want to use slash-dot to make a replacement but you want to make more than one pass through the expression. If we use *slash-slash-dot, //.*, *Mathematica* will keep passing through the expression over and over again, making substitutions with each pass, until the expression ceases to change. In Example 4.2.9, we use a replacement rule that will turn any product into a sum and illustrate what happens if we use slash-dot versus slash-slash-dot.

Example 4.2.9

```
In[58]:= (* turning products to sums *)
    a b c /. x_ y_ → x + y
    a b c //. x_ y_ → x + y

Out[58]= a + b c

Out[59]= a + b + c
```

Notice first the structure of our replacement rule **x_ y_ → x y**. The left-hand side, **x_ y_**, is a *pattern* that, in this case, stands for the product of any two things. In general, a pattern is any expression that contains underscores, or "blanks." So **x_**, **f[x_]**, **a_+b_** are all patterns. Even a single blank all by itself, **_**, is a pattern. In a pattern, blanks can stand for anything. The difference between **x_** and **_**, is that the first pattern stands for anything, but we have named the anything **x**, whereas the second pattern also stands for anything, but we have not given it a name. By using **x_** and giving the anything a name, we can then use the name on the right-hand side of a replacement rule. Thus in our replacement rule, the pattern **x_ y_** stands for the product of any two things and furthermore, we have named those two things **x** and **y** so that we can use those names to form the right-hand side of the replacement rule, namely, **x+y**. When this replacement rule is applied to an expression using the slash-dot function, *Mathematica* looks for subexpressions that match the pattern of the left-hand side of the replacement rule. When it finds a match it replaces the subexpression according to the rule.

Next, notice that if we use slash-dot and only make one pass through the expression we end up with *a+b c*, whereas if we use slash-slash-dot and repeatedly pass through the expression until it no longer changes, we obtain *a+b+c*.

It is quite easy using slash-slash-dot to create an *infinite loop*, that is, the process of substitutions will go on forever and never end. If this happens your entire hard drive will be filled with nonsense and your computer will be ruined! Just kidding—*Mathematica* will usually stop the process with a warning. But if not, you can abort the calculation.

It is possible to use a replacement rule and specify exactly how many passes through the expression should be made. In the Quiz, we include a question exploring this topic.

4.3 Working with Lists

By now it should be obvious that lists are quite important in *Mathematica*. There are quite a few functions that can be used to operate on lists and we'll describe a handful of them in this section.

We have already seen that if **x** is the name of a list, then **x[[n]]** will give the nth element of the list. A nice variation on this is that **x[[-n]]** will give the nth element from the end of the list. Picking out specific elements can also be done with the **Part** function, for which the double square bracket notation, **[[ ]]**, is really shorthand. In Example 4.3.1, we give a couple of examples. We first define a list of even integers and name it **evens**. Next we use the **Part** function (or its abbreviation using double square brackets) to extract certain elements.

Example 4.3.1

```
In[36]:= (* picking certain elements out of a list *)
     evens = {0, 2, 4, 6, 8, 10, 12, 14, 16};
     evens[[3]]
     Part[evens, 3]
     evens[[-2]]
     Part[evens, -2]
```

Out[37]= 4

Out[38]= 4

Out[39]= 14

Out[40]= 14

If we want to extract more than a single element from a list we can use the **Take** function. The expression **Take[evens, n]** will give the first n elements of the list. We can also use **-n** to return the last n elements. Example 4.3.2 also shows how replacing **n** with the list {**n, m**} will return elements n through m.

Example 4.3.2

```
In[19]:= (* extracting consecutive elements from a list *)
    Take[evens, 2]
    Take[evens, -3]
    Take[evens, {2, 4}]
```

Out[19]= {0, 2}

Out[20]= {12, 14, 16}

Out[21]= {2, 4, 6}

The opposite of **Take** is **Drop**. Using **Drop** we can take all the elements of a list except certain elements. We illustrate this in Example 4.3.3. It is important to realize that **Take** and **Drop** do NOT change the value of the list they operate on. In the above examples, **evens** is still the same after using **Take** or **Drop**. These functions simply *return* lists that are extracted from **evens** while leaving **evens** alone. If we wanted to remove a certain element from a list we could use **Drop** to do it and then rename the list with the results of **Drop**. For example, after we evaluate **evens=Drop[evens, 1]** the value of evens will be {2, 4, 6, 8, 10, 12, 14}, the initial element of 0 having been dropped.

Example 4.3.3

```
In[29]:= (* using Drop to remove elements from a list *)
    evens
    Drop[evens, 3]
    Drop[evens, -2]
    Drop[evens, {2, 3}]
```

Out[29]= {0, 2, 4, 6, 8, 10, 12, 14, 16}

Out[30]= {6, 8, 10, 12, 14, 16}

Out[31]= {0, 2, 4, 6, 8, 10, 12}

Out[32]= {0, 6, 8, 10, 12, 14, 16}

You definitely need to look at the Help Files pages for **Take** and **Drop** as there are several variations on the way they can be used that are quite important.

Two other important functions are **Sort** and **Reverse** which do the obvious thing to a list. Example 4.3.4 shows their use. Note that we first use **RandomInteger[10]** together with **Table** to produce a list of 10 randomly chosen integers that each lie between 0 and 10.

Example 4.3.4

```
In[39]:= (* sorting and reversing a list *)
    myList = Table[RandomInteger[10], {10}]
    Sort[myList]
    Reverse[Sort[myList]]

Out[39]= {6, 0, 5, 4, 3, 10, 10, 9, 5, 3}

Out[40]= {0, 3, 3, 4, 5, 5, 6, 9, 10, 10}

Out[41]= {10, 10, 9, 6, 5, 5, 4, 3, 3, 0}
```

If we think of lists as sets, then we might want to find the intersection or union of two or more sets. In Example 4.3.5, we use the functions **Intersection** and **Union** to do this. Each function takes any number of lists as arguments, separated by commas. In the fist line, for example, we enter three sets as arguments to **Intersection**. The only elements that are common to all three sets are 3 and 4, and thus {3, 4} is returned. Similarly, **Union** will give all the elements that appear in at least one of the sets. So the union of the same three sets is now the integers from 1 to 6. Both of these functions return their results in sorted order.

Example 4.3.5

```
In[58]:= (* Intersection and Union treat lists as sets *)
    Intersection[{1, 2, 3, 4}, {2, 3, 4, 5}, {3, 4, 5, 6}]
    Union[{1, 2, 3, 4}, {2, 3, 4, 5}, {3, 4, 5, 6}]
    Union[{1, 4, 0, 2, 6, 6, 4, 7, 2, -2, 0, 1, 2, 7}]

Out[58]= {3, 4}

Out[59]= {1, 2, 3, 4, 5, 6}

Out[60]= {-2, 0, 1, 2, 4, 6, 7}
```

A tricky way to use either of these functions is to pass in only one list. In this case, the list will be returned sorted and with duplicate elements removed. In the third line above, the input list has 14 elements, but some of them, like 6, are repeated. After applying **Union**, the duplicates are removed and the list is sorted.

Let's look at a useful example. Suppose we are given a list of numbers named **data** and we want to remove the two lowest values. If we don't care about the order of the elements in the list, a simple solution is shown in Example 4.3.6. We simply sort the data, and drop the first two elements from the sorted list. Notice that we use **RandomReal[ ]** together with **Table** to produce a list of randomly chosen real numbers between 0 and 1. Looking carefully at the first list we see that .00153465 and .0301557 are the two smallest elements and they do not appear in the second list.

Example 4.3.6

```
In[3]:= (* dropping two lowest values from a list *)
    (* order of list is lost *)
    data = Table[RandomReal[], {10}]
    Drop[Sort[data], 2]

Out[3]= {0.142245, 0.543427, 0.248698,
    0.282653, 0.471051, 0.526607, 0.849228,
    0.401546, 0.0118537, 0.485391}

Out[4]= {0.248698, 0.282653, 0.401546, 0.471051,
    0.485391, 0.526607, 0.543427, 0.849228}
```

If we want to drop the two lowest elements from a list but still maintain the order of the list, the problem is a bit harder. If we sort the list and then drop the lowest two, how will we be able to put the list back into order? There is a way to do this, but another strategy would be to never sort the list in the first place. We need to find the two smallest elements and drop them. The next example solves the problem.

Example 4.3.7

```
In[5]:= (* removing the two lowest elements from
    a list and maintaining order of list *)
    data = Table[RandomInteger[10], {15}]
    minPlaces = Position[data, Min[data]]
    temp = Drop[data, minPlaces[[1]]]
    minPlaces = Position[temp, Min[temp]]
    Drop[temp, minPlaces[[1]]]

Out[5]= {5, 10, 3, 9, 3, 0, 0, 8, 4, 8, 6, 8, 8, 1, 8}

Out[6]= {{6}, {7}}

Out[7]= {5, 10, 3, 9, 3, 0, 8, 4, 8, 6, 8, 8, 1, 8}

Out[8]= {{6}}

Out[9]= {5, 10, 3, 9, 3, 8, 4, 8, 6, 8, 8, 1, 8}
```

The key to Example 4.3.7 is to use the **Position** function. In general, **Position[**list, x**]** will return a list of all the positions in which x occurs. Let's take a careful look at Example 4.3.7 to see how it works. We begin by forming a list of 15 randomly chosen integers between 0 and 10. We name this list **data**. Next we use **Position[data, Min[data]]**. Given any list, **Min** will give the minimum value in the list, and **Max** will give the maximum value. So **Position[data, Min[data]]** will give a list of all the places the minimum occurs. In this case the minimum is 0 and appears in positions 6 and 7. Therefore, **minPlaces[[1]]** will be the first place where the minimum occurs. Next, **Drop[data, minPlaces[[1]]]** will remove the first occurrence of the minimum value. Then, in the next two lines we repeat the process. We use the name **temp** to stand for the intermediate list obtained after we remove one element.

We have already mentioned that most functions in *Mathematica* are *listable*. That is, if we enter a list as an argument to a function, we'll get back the list obtained by letting the function act on each element of the original list. For example, **Abs[{x,y,z}]** will return {**Abs[x], Abs[y], Abs[z]**}. We say that the function *threads* over the list.

Another way to accomplish the same thing is to use the **Map** function. In general, **Map[**f, expr**]** will apply f to the first *level* of parts in expr. Example 4.3.8 illustrates the use of **Map**.

Example 4.3.8

```
In[136]:= (* Map applies the function to the
          first level of parts in each expression *)
       Map[Abs, {x, y, z}]
       Map[Abs, {x, y, {z, w}}]
       Map[Sin, x + y x]
       Map[f, x y z]

Out[136]= {Abs[x], Abs[y], Abs[z]}

Out[137]= {Abs[x], Abs[y], {Abs[z], Abs[w]}}

Out[138]= Sin[x] + Sin[x y]

Out[139]= f[x] f[y] f[z]
```

When *expr* is a list, its first level of parts are its elements. Thus in the first line in Example 4.3.8 we get the list {**Abs[x], Abs[y], Abz[z]**}, the same as what we would get from **Abs[{x, y, z}]**. In the second line, **Map** will give {**Abs[x], Abs[y], Abs[{z, w}]**} but since **Abs** is listable, the final element of the list is rewritten as {**Abs[z], Abs[w]**}. On the third line we obtain **Sin[x]+Sin[y x]** because x and x y are the first level of parts of the expression $x + x\,y$. Finally, in the last line, the first level of parts in a product are the factors. Hence f is applied to each of the three factors.

We'll see places later in this book where it's handy to use the **Map** function. After we talk about functions in Chap. 5, and especially what are called *pure* functions, we'll see another way to use the **Map** function. We also need to learn more about functions before we can understand the **Select** function. **Select** is an absolutely incredible function that can be used to select elements from a list that satisfy certain properties.

We can also combine lists of the same size with operations that would normally be used to combine just two numbers (or variables). Because the lists are the same size, the operation is just applied to corresponding elements. We give a few examples in Example 4.3.9.

Example 4.3.9

```
In[51]:= (* combining lists of the same size *)
         {1, 2, 3, 4} + {0, 0, -1, 2}
         {1, 2, 3, 4} * {0, 0, -1, 2}
         {a, b, c, d}^{-1, 0, 1, 2}
         {a, b, c} / {e, f, g}^{2, 2, 2}
```

$$\text{Out[51]} = \{1,\ 2,\ 2,\ 6\}$$

$$\text{Out[52]} = \{0,\ 0,\ -3,\ 8\}$$

$$\text{Out[53]} = \left\{\frac{1}{a},\ 1,\ c,\ d^2\right\}$$

$$\text{Out[54]} = \left\{\frac{a}{e^2},\ \frac{b}{f^2},\ \frac{c}{g^2}\right\}$$

In the last example, if the lists were not of the same size, *Mathematica* would object and issue a warning. However, if one of the lists is just a single number or variable, that is a *scalar*, then the scalar will be treated as a list of the correct size all of whose elements are the same. We give a few examples of this below. Understanding Examples 4.3.9 and 4.3.10 can really streamline how we handle expressions in lots of cases. Notice that in the last line of Example 4.3.9, the exponentiation has a higher precedence than division, so is done first.

Example 4.3.10

```
In[47]:= (* combining a list with a scalar*)
         {a, b, c} + 3
         {a, b, c} * x
         {a, b, c} ^ 4
         {a, b, c, d} / 5

Out[47]= {3 + a, 3 + b, 3 + c}

Out[48]= {a x, b x, c x}

Out[49]= {a⁴, b⁴, c⁴}

Out[50]= {a/5, b/5, c/5, d/5}
```

There are quite a few other functions that can be applied to lists such as **Append, AppendTo, Flatten, Join**, and **Partition** as well as other interesting ways in which lists can be combined. Rather than trying to explain all the possibilities here, we'll introduce more ways to deal with lists in later sections.

4.4 Sums and Products

Suppose we want to find the sum of the squares of the first 1000 positive integers. One way to do this would be to place the summands in a list and then use the **Total** function. In general, **Total** takes a list as its argument and will return the sum of all the elements in the list. Here is an example, where we have used only the first 10 positive integers.

Example 4.4.1

```
In[149]:= (* using Total to add the squares of
          the first 10 positive integers *)
          squares = Table[i², {i, 1, 10}]
          Total[squares]

Out[149]= {1, 4, 9, 16, 25, 36, 49, 64, 81, 100}

Out[150]= 385
```

In the above example, we first formed a list with the summands we wanted to add and then used **Total**. If we use **Sum** instead, we do not need to make the intermediate step of forming the list of summands. In Example 4.4.2 we recompute the sums of the squares of the first 10 positive integers using **Sum**. We provide two arguments to **Sum**. The first is the formula for the ith summand of the sum and the second is the familiar indexing list used to control i, the index of summation. The syntax of **Sum** is very much like the syntax of **Table**, and in fact, using **Sum** is equivalent to using **Table** followed by **Total**.

Example 4.4.2

In[152]:= (* using Sum to add the squares of the
         first 10 positive integers *)
       Sum$\left[i^2, \{i, 1, 10\}\right]$

Out[152]= 385

    Of course, if we wanted to add up the squares of the first 10000 positive integers it would be easier and faster to simply use the closed formula for the sum. You might remember that there are really neat closed formulas for the sums of the kth powers of the first n positive integers. *Mathematica* is aware of these formulas and can apply them when using **Simplify** as the next example shows.

Example 4.4.3

In[154]:= (* Simplify is aware of many summation
         formulas *)
       Simplify$\left[\text{Sum}\left[i^2, \{i, 1, n\}\right]\right]$

Out[154]= $\frac{1}{6}$ n (1 + n) (1 + 2 n)

    Finally, there is a function called **Product**, which works much like **Sum** except that it forms the product of the given terms rather than their sum. We close this section with an example giving the product of the first 10 primes.

Example 4.4.4

In[160]:= (* using Product to multiply the first
        10 primes together *)
        Product[Prime[i], {i, 1, 10}]

Out[160]= 6 469 693 230

In[161]:= (* factoring the previous result *)
        FactorInteger[%]

Out[161]= {{2, 1}, {3, 1}, {5, 1}, {7, 1}, {11, 1},
        {13, 1}, {17, 1}, {19, 1}, {23, 1}, {29, 1}}

4.5 Matrices

A *matrix* is an array of numbers (or other objects) arranged in rows and columns. In *Mathematica* matrices are expressed as a list of the rows, each of which, of course, is a list itself. Thus a matrix is a list of lists. If a matrix has *n* rows and *m* columns, we call it an *n* by *m* matrix. The entry in the *i*th row and *j*th column is called the *i*, *j* entry. In Example 4.5.1, we define a 3 by 4 matrix as a list of three rows, each a list of 4 elements. We then use the function **MatrixForm** to print out the array in rows and columns. **MatrixForm** will also enclose the matrix in large parenthesis, which is common practice for writing matrices.

Example 4.5.1

In[191]:= (* a matrix is a list of rows *)
        B = {{1, 2, 3, 4}, {a, b, c, d}, {0, 0, 1, 0}}
        MatrixForm[B]

Out[191]= {{1, 2, 3, 4}, {a, b, c, d}, {0, 0, 1, 0}}

Out[192]//MatrixForm=

$$\begin{pmatrix} 1 & 2 & 3 & 4 \\ a & b & c & d \\ 0 & 0 & 1 & 0 \end{pmatrix}$$

We have already seen how to use **Table** to create lists. In fact, we can use it to create matrices as seen in Example 4.5.2. All we need to do is give two indexing lists to **Table**. The first will control the rows and the second will control the columns. In Example 4.5.2, the *i*, *j* entry is the abstract expression $b_{\{i, j\}}$.

Example 4.5.2

In[17]:= (* using Table with multiple indexing lists *)
    m = Table[
        b_{row, column},
        {row, 1, 4, 1}, {column, 1, 2, 1}
        ];
    MatrixForm[m]

Out[18]//MatrixForm=

$$\begin{pmatrix} b_{\{1,1\}} & b_{\{1,2\}} \\ b_{\{2,1\}} & b_{\{2,2\}} \\ b_{\{3,1\}} & b_{\{3,2\}} \\ b_{\{4,1\}} & b_{\{4,2\}} \end{pmatrix}$$

Actually, what we have just called a matrix should more properly be called a *two-dimensional* matrix. We can extend the notion to any dimension. Going down a dimension, a one-dimensional matrix is simply a list. Going up a dimension, a three-dimensional matrix would be a list of two-dimensional matrices, all the same size (i.e., having the same number of rows and columns). In general, an *n*-dimensional matrix is a list of $(n-1)$-dimensional matrices, all of the same size. For two-dimensional matrices we usually won't mention the dimension and simply use the word "matrix."

It is worth noting that we can use **Table** to create matrices of any dimension. We simply need to add as many indexing lists as there are dimensions.

Example 4.5.3

In[15]:= (* multiplying matrices *)
    MatrixForm[A = {{1, -1}, {2, 0}, {x, y}, {a, a}}]
    MatrixForm[B.A]

Out[15]//MatrixForm=

$$\begin{pmatrix} 1 & -1 \\ 2 & 0 \\ x & y \\ a & a \end{pmatrix}$$

Out[16]//MatrixForm=

$$\begin{pmatrix} 5+4\,a+3\,x & -1+4\,a+3\,y \\ a+2\,b+a\,d+c\,x & -a+a\,d+c\,y \\ x & y \end{pmatrix}$$

If you have had a course in linear algebra, you know that an *n* by *m* matrix can be multiplied times an *m* by *r* matrix. In Example 4.5.3, we define a new matrix **A** and form the product **B.A** with the matrix of Example 4.5.1. Notice that the period is used as the symbol for matrix multiplication.

Matrices are very important in many branches of mathematics and *Mathematica* has quite a few functions that deal with matrices. We'll be introducing them as we go.

4.6 Find Out More

Here are a few nice tutorials from the Help Files that relate to the topics of this chapter:

- tutorial/TransformingAlgebraicExpressions and tutorial/Simplifying-AlgebraicExpressions—contain lots of good examples related to simplifying expressions. (These are both part of the larger tutorial tutorial/Algebraic-CalculationsOverview.)
- tutorial/ApplyingTransformationRules—a great tutorial about the slash-dot replacement function. (This tutorial is part of the larger tutorial/TransformationRulesAndDefinitionsOverview.)
- tutorial/PatternsOverview—excellent discussion of patterns and their use in replacement rules.
- tutorial/ListsOverview—a comprehensive overview of lists.
- tutorial/ApplyingFunctionsToPartsOfExpressions—a nice discussion on **Map**.

Quiz

1. Simplify $\sqrt{\frac{w^3 x^{-2} y^5}{w^5 x z^3}}$.
2. Use **Simplify** to find a formula for the sum of the cubes of the positive integers from 1 to n.
3. Rewrite $\sin(4\theta)$ in terms of trigonometric functions of θ alone. (That is, $\sin\theta$, or $\cos\theta$, for example, may appear in the rewritten expression, while $\tan(2\theta)$ or $\cos(3\theta)$ may not.)
4. Simplify $-4 \sin^3 x + 4 \cos x \sin^2 x + 3 \sin x - \cos x$.
5. Use **Table** and **/.** to make a list of the polynomials obtained from $x^5 - 3x^2 + 6$ by replacing x with $x + k$ for $k = -4, -3, \ldots, 3, 4$.

6. Let $p_3(n)$ be the coefficient of x^3 in the polynomial $(x^2 + x + 1)^n$. Make a table of $p_3(n)$ for $n = 2, 3, \ldots, 25$.

7. Write replacement rules using patterns that will change $\log(xy)$ to $\log x + \log y$ and $\log \frac{x}{y}$ to $\log x - \log y$. Apply the rules to $\log \frac{xyz}{w}$ so as to produce $\log x + \log y + \log z - \log w$.

8. The construction **ReplaceRepeated[expression, rules, MaxIterations → k]**, will perform the substitution rules on **expression** a total of **k** times. Use this, starting with the expression **x**, to obtain

$$1 + \cfrac{1}{1 + \cfrac{1}{1 + \cfrac{1}{1 + \frac{1}{x}}}}$$

9. Create a list of 20 randomly chosen real numbers and then figure out how to remove the smallest and largest numbers, keeping the order of the list otherwise intact.

10. If **data** is a list of at least 10 elements, figure out how to trade the fourth and fifth elements of **data**.

11. A famous *infinite* product, discovered in 1665 by the English mathematician John Wallis, is

$$\frac{\pi}{2} = \frac{2}{1} \cdot \frac{2}{3} \cdot \frac{4}{3} \cdot \frac{4}{5} \cdot \frac{6}{5} \cdot \frac{6}{7} \cdots$$

Use **Product** to compute the product of the first 10, 100, and 1000 factors on the right-hand side of the equation. Do these *partial* products come close to $\frac{\pi}{2}$? What happens, if instead of writing 10, 100, or 1000 as the upper limit of the indexing list in **Product**, you write **Infinity**?

CHAPTER 5

Functions

Even though *Mathematica* comes with thousands of predefined functions, it is extremely important to know how to define our own functions. While a simple function might be defined in a single line of code, more complicated functions might take dozens of lines assembled together in a **Module**. In this chapter we'll see how to define our own functions. We'll also learn the basics of elementary *programming* which can be used to turn any mathematical algorithm into a working *Mathematica* function.

5.1 Defining Functions

Sometimes we may be using the same function over and over again and it can be handy to give the function a name. For example, suppose we were computing things related to home mortgages and repeatedly found ourselves computing the monthly payment on a loan of $400 000 amortized over 30 years at an annual interest rate of r. Imagine that we keep changing the interest rate r and have to keep retyping the formula for the monthly payment. To save on all that typing, we can create our own function to compute the payment as seen in the following example.

Mathematica Demystified

Example 5.1.1

In[56]:= `(* defining our own function *)`
`    (* the monthly payment on a $400000 loan`
`     over 30 years at interest rate r *)`

$$\texttt{payment[r\_]} := \frac{400\,000\,\frac{r}{12}\left(1+\frac{r}{12}\right)^{360}}{\left(1+\frac{r}{12}\right)^{360}-1}$$

On the left-hand side of the definition we have named the function **payment** and used as its argument the pattern **r_**. (In fact, the entire left-hand side, **payment[r_]** is the pattern—but more about that in just a moment.) Because of the use of the blank, **r_** can stand for anything, and in particular, any interest rate. On the right-hand side we give the formula for the monthly payment as a function of the interest rate r. Notice that we do not use a blank on the right-hand side.

Between the two sides of the definition we use not an equals sign, but colon-equals. The equals sign is actually shorthand for the function **Set** while the colon-equals is shorthand for the function **SetDelayed**. We have already seen the use of the equals sign many times. Using *lhs* = *rhs* will cause *rhs* to be evaluated and, forever after, this result will be substituted for *lhs* wherever it occurs. This is exactly what we want when we give a variable a value, or name something that we have created. Using colon-equals almost gives the same thing, but not quite. Instead, whenever *lhs* occurs it is replaced by *rhs* and then it is evaluated. This is a pretty subtle difference and we'll give some examples to try and make the distinction clear. But, generally speaking, you'll almost always be in good shape if you always use colon-equals in the functions you define.

Finally, one more point about the function definition. When we define a function like **g[z_]:=z+π**, all we are telling *Mathematica* is to use the replacement rule **g[z_]** $\rightarrow$ **z+π** whenever possible. So we see that it is really the entire left-hand side of the definition that serves as the pattern, not just the argument **z_**.

Having defined the function **payment** we could use it, for example, to compute the monthly payments for loans with different interest rates ranging from 5% to 8%. We do this in Example 5.1.2.

Example 5.1.2

In[55]:= `(* loan payments for interest rates`
`     ranging from 5 to 8 percent *)`
`    Table[payment[r], {r, .05, .08, .005}]`

Out[55]= `{2147.29, 2271.16, 2398.2,`
`    2528.27, 2661.21, 2796.86, 2935.06}`

Notice that when we use the function **payment** we do not use the blank after its argument. Without the blank, **payment[r]** will match the pattern used to define it and hence be replaced by the formula for the payment. Also note that we entered the interest rate of 5% as .05, and that we made the interest rate advance in increments of half a percent. The monthly payments range from around $2150 to around $2935, a difference of almost $800. Clearly it pays to shop around for the lowest interest rate available.

Suppose that after using the **payment** function for awhile, we find that sometimes we need to figure the monthly payments for other loan amounts. No problem! We'll simply rewrite the function to take two arguments: the interest rate and the loan amount, or *principal*. In Example 5.1.3, we have enlarged the pattern on the left-hand side to include two (named) blanks, one for each argument, and the right-hand side is a function of both P and r. It doesn't matter which argument we list first. But, of course, now that we have made a choice, we need to have the principal first when we call the function.

Example 5.1.3

In[60]:= (* defining a function with two arguments *)
    (* the monthly payment on a 30-year loan
    with principal P and interest rate r *)

$$payment[P\_, r\_] := \frac{P \, \frac{r}{12} \left(1 + \frac{r}{12}\right)^{360}}{\left(1 + \frac{r}{12}\right)^{360} - 1}$$

Let's use our function to make a table of monthly payments corresponding to different initial loan amounts and different interest rates. Example 5.1.4 shows how to do this. To understand how it works, let's start with the **Table** function. Notice that we are using **Table** with two indexing lists, one for P and another for r, so that we create a two-dimensional array of monthly payments. Since the indexing list for P appears first, rows correspond to the principal and the columns correspond to the interest rate.

Next, we have used **TableForm** to print the table in a nice tableaux. **Table-Form** is nice to use because it has the option **TableHeadings** that we have used here to include the row of interest rates as well as the column of loan amounts. The vertical and horizontal dividing lines are automatically included when using **TableHeadings**. Alternatively, we could have used **Grid** to layout the table nicely. **Grid** can be used to place all kinds of objects in a gridlike format and so is similar to **TableForm**. But **Grid** is more general in the sense that it can be used to arrange more complicated objects in sophisticated ways.

Example 5.1.4

```
In[202]:= (* making a table of monthly payments for
        different interest rates and different loan
        amounts *)
     mortgageTable =
      TableForm[
       Table[
        payment[P, r],
        {P, 300000, 500000, 50000},
        {r, .06, .08, .005}
        ],
       TableHeadings → {
         Table[P, {P, 300, 500, 50}],
         Table[r, {r, .06, .08, .005}]
         }
       ]
```

Out[202]//TableForm=

	0.06	0.065	0.07	0.075	0.08
300	1798.65	1896.2	1995.91	2097.64	2201.29
350	2098.43	2212.24	2328.56	2447.25	2568.18
400	2398.2	2528.27	2661.21	2796.86	2935.06
450	2697.98	2844.31	2993.86	3146.47	3301.94
500	2997.75	3160.34	3326.51	3496.07	3668.82

If we want to add textual labels to our mortgage payment table it would be hard to do with **TableForm** and perhaps easier with **Grid**. We can also combine the two functions to build up the final layout. For example, having already named the layout in Example 5.1.4 **mortgageTable**, we could now use it as one element in a grid with the other elements being text labels. In Example 5.1.5, we use **Grid** to layout two elements one on top of the other: the text label and the **mortgageTable**. Notice the use of **Text** and **Style** (described earlier in Chap. 2) to create the label.

There is a nice guide in the Help Files that describes how to layout tables. We'll point you to it in the Getting Help section of this chapter.

Example 5.1.5

```
In[203]:= (* adding a label to the table *)
    label = Text[
      Style[
        "Mortgage payments per interest rate and
          initial loan balance.", 18]];
    Grid[{{label}, {mortgageTable}}]
```

Mortgage payments per interest rate and initial loan balance.

	0.06	0.065	0.07	0.075	0.08
300	1798.65	1896.2	1995.91	2097.64	2201.29
350	2098.43	2212.24	2328.56	2447.25	2568.18
400	2398.2	2528.27	2661.21	2796.86	2935.06
450	2697.98	2844.31	2993.86	3146.47	3301.94
500	2997.75	3160.34	3326.51	3496.07	3668.82

Out[204]=

As the final example in this section, let's define a function that will give the n-th Fibonnaci number. Remember that the Fibonnaci numbers are $0, 1, 2, 3, 5, 8, 13, \ldots$ where each number in the sequence is defined as the sum of the previous two. Letting the first and second numbers be zero and one respectively gets the ball rolling.[1] Defining the nth Fibonnaci number as the sum of the previous two is an example of a *recursive* definition; we use the function to define itself! It's not quite circular though because we define each number in term of previous numbers and eventually we bottom out at the first and second number which have solid, non-recursive definitions. The next example defines the function.

Example 5.1.6

```
In[37]:= (* recursive definition of Fibonnaci numbers *)
    fib[0] = 0;
    fib[1] = 1;
    fib[n_] := fib[n - 1] + fib[n - 2]

In[50]:= (* computing the 30th Fibonnaci number *)
    Timing[fib[30]]

Out[50]= {4.30065, 832 040}
```

Notice that we use the ordinary equals sign to define **fib[0]** and **fib[1]** and then the colon-equals sign to define **fib[n_]** in terms of **fib[n − 1]** and **fib[n − 2]**. In the

---

[1]Somewhat arbitrarily, we'll call 0 the 0th Fibonnaci number, 1 the first Fibonnaci number, 2 the second, and so on.

second input cell where we compute **fib[30]** (and time how long it takes) *Mathematica* must first replace **fib[30]** by the expression **fib[29]+fib[28]** and then evaluate this expression. But to evaluate **fib[29]** *Mathematica* must replace **fib[29]** with **fib[28]+fib[27]** and then evaluate that, and so on. Clearly this mushrooming process is going to take a lot of time.[2] We'll see a much faster way to compute **fib[30]** a little later in the chapter.

If you think about it, using an equals sign instead of a colon-equals sign in this example will not work. If we use an equals sign *Mathematica* will try to evaluate the right-hand side of the definition and then assign that value to the left-hand side. But in order to evaluate the right-hand side it needs to know what the definition is—the one that we are defining right now! Try it out and see what happens.

When we define our own functions we are creating symbols that name the function and we should follow the same advice that applies to variable names or anything else that we name. Since all functions in *Mathematica* start with capital letters a good practice for us to follow is to create function names that begin with lowercase letters. Secondly, we should try to use descriptive names. Using **fib** or even **fibonnaci** is much better than just using **f**. Of course, it is nice to use shorter names, so we won't always spell things out completely. Finally, we should avoid names that are extremely close to existing names.

5.2 Pure Functions

When we define a function such as **g[x_]:=x Sin[x]** it is usually because we want to use the function over and over. Defining it this way gives it a name (in this case **g**) which makes it easy to refer to.

On the other hand, there are times when we might want to use a function only once, in which case it might not even be worth naming the function. In this setting we may want to use *pure functions*. We'll give four examples in this section that use pure functions.

Suppose for example that we want to plot a list of complex numbers in the plane. Recall that the complex number $x + iy$ corresponds to the point in the plane with Cartesian coordinates (x, y). If the numbers were given by their coordinates, then it would be a simple matter to use **ListPlot** to plot them. So, we need a function that can take a complex number and turn it into the corresponding pair of coordinates. If we had a whole list of complex numbers we could use **Map** to apply this function

---

[2]Try entering **Trace[fib[3]]**. *Mathematica* will print out all the intermediate steps it follows to compute **fib[3]**. It shouldn't be too hard for you to figure out what is going on, but this is getting a little more advanced then we should be. If you are really interested you can read about **Trace** in the Help Files. Then enter **Trace[fib[10]]** to just get a hint of how many steps need to be carried out to compute **fib[30]**!

to every number in the list. There is a simple solution using pure functions that is shown in Example 5.2.1.

Example 5.2.1

```
In[212]:= (* mapping a pure function onto a set of
       complex numbers to turn them into points *)
     data = Table[RandomComplex[], {6}]
     Map[{Re[#], Im[#]} &, data]

Out[212]= {0.783791 + 0.471847 i, 0.523661 + 0.901026 i,
       0.73807 + 0.841179 i, 0.889937 + 0.929982 i,
       0.314625 + 0.147453 i, 0.808857 + 0.894805 i}

Out[213]= {{0.783791, 0.471847}, {0.523661, 0.901026},
       {0.73807, 0.841179}, {0.889937, 0.929982},
       {0.314625, 0.147453}, {0.808857, 0.894805}}
```

The first thing we do in Example 5.2.1 is create a list of six complex numbers by using the **Table** function and the function **RandomComplex[ ]** which will return a randomly chosen complex number. We name the list **data** and it is displayed in the first output cell. Next we use **Map** to apply a function to every element of **data**, converting each one to a list of its x and y coordinates. The first argument of **Map** is the function we want to apply and this is where we hit new territory. Instead of placing the name of a function here, we just describe the function using the syntax of pure functions.

Just as we are not going to take the trouble to name this function, we also do not name its arguments! Instead, the arguments are specified by "slot numbers" #n. If there is only one argument it is represented by #. But if there are multiple arguments then #1 will refer to the first, #2 to the second, and so on. The expression {**Re[#], Im[#]**} will take the single argument and form a list whose first element is the real part of the argument and whose second element is the imaginary part of the argument. This is exactly what we want! Finally, the ampersand, &, is extremely important. It tells *Mathematica* that the expression is defining a pure function.

As a second example, suppose we have a data set that consists of a list of elements, where each element is itself a list of the following items: name, age, and telephone number. Suppose we want to extract all the ages from the data set. That is, we want to extract the second element of each element of the big list. Example 5.2.2 does this by using the pure function **#[[2]]&**, which returns the second element of the argument. Example 5.2.2 also illustrates the alternative syntax *f/@expr* that can be used instead of **Map[***expr, f***]**.

Example 5.2.2

```
In[21]:= (* using a pure function to extract the
         ages from the data set *)
      directory = {
          {"Alice", 12, 3 451 231},
          {"Bob", 11, 3 456 785},
          {"Carol", 14, 3 455 674},
          {"Dylan", 13, 3 452 390}
          };
      #[[2]] & /@ directory
Out[22]= {12, 11, 14, 13}
```

Of course, there are other ways to accomplish the same thing as Example 5.2.2 without using pure function. We might, for example, use

Table[directory[[k,2]], {k,1,Length[directory]]}

But Example 5.2.2 is simpler.

Suppose that we want to sort the entries of **directory** in order of age. If we use **Sort[directory]** the triples will be sorted on their first element. This would arrange the data alphabetically by name. Fortunately, **Sort** accepts an option that allows us to define what it means for one item to be less than another, and **Sort** will then use our rule. Example 5.2.3 illustrates this. The sorting rule we provide is **#1[[2]]<#2[[2]]&**. **Sort** uses this rule to compare two items which are referred to here by their slot numbers **#1** and **#2**. If the second element of the first argument is less than the second element of the second argument then our rule will evaluate to **True**, and **Sort** will order the two items accordingly.

Example 5.2.3

```
In[228]:= (* sorting the directory on age by providing
         our own sorting rule to Sort *)
      Sort[directory, #1[[2]] < #2[[2]] &]
Out[228]= {{Bob, 11, 5 323 452}, {Alice, 12, 5 321 234},
         {Dylan, 13, 5 321 254}, {Carol, 14, 5 321 003}}
```

As a final example illustrating the use of pure functions, we describe the **Select** function. This function is used to extract elements from a list that meet certain requirements. For example, suppose we want to extract all the people from the

directory of Example 5.2.2 who are older than 12. Example 5.2.4 does just what we want by using the **Select** function.

Example 5.2.4

```
In[26]:= (* selecting all people in directory that
         are older than 12 *)
      Select[directory, #[[2]] > 12 &]
Out[26]= {{Carol, 14, 3 455 674}, {Dylan, 13, 3 452 390}}
```

   In general, **Select** takes two arguments. The first is the list of data and the second is a criteria. All data elements that satisfy the criteria are returned. Here we have used a pure function to express the criteria. If the second entry of an element is more than 12, the criteria will evaluate to **True** and the element will be extracted.

   There are other ways to do the above examples without using pure functions, but using pure functions is often quite simple and elegant. So it is definitely worth understanding pure functions for those occasions when it makes sense to use them. Finally, a very common error that will almost certainly give you grief, is *leaving off the ampersand in a pure function*. Try to watch out for that!

5.3 Elementary Programming

So far we have used *Mathematica* primarily to perform calculations or produce neat graphics. But as we try to perform more and more complicated calculations we will need to write simple *programs* to carry out whatever algorithm is needed to complete the calculation. The essence of programming consists of being able to give the computer a set of instructions to carry out, one after the other, and included in these, logical instructions that can control the *flow of execution*, that is, the order in which the instructions are performed.

   One of the simplest and most important constructions that can alter the flow of execution is the *If-Then-Else* statement. Every programming language provides this construction. In *Mathematica* the **If** function provides this essential ingredient. The syntax for the **If** function is

$$\textbf{If}[\,condition,\, t,\, f\,]$$

When the **If** function is evaluated, the **condition** is evaluated to see if it is **True** or **False**. If it is **True**, then the expression t is returned, otherwise the expression f is returned. In Example 5.3.1, we define a function named **f** using **If**. If the condition is true, that is $x > 0$, then the second argument of the **If** function, namely **Sin[x]**, is returned. If it is not the case that $x > 0$, then 0 is returned. Thus **f[x]:=Sin[x]** for positive x and **f[x]:=0** otherwise.

Example 5.3.1

In[281]:= `(* using If to define a function *)`
    `f[x_] := If[x > 0, Sin[x], 0]`

Here is a second example where the function **g[n]** defined in Example 5.3.2 returns $n/2$ if n is even and $3n+1$ if n is odd. (We intend to only apply this function to integers.) The condition uses the **Mod** function. In general, **Mod**[x,y] gives the remainder when x is divided by y. Thus an integer n is even if and only if **Mod**[n, **2**] is zero. Notice that in our condition, or test, we use double equals signs. A single equals sign is used for assignment: the right-hand side is assigned to the left-hand side. But a double equals sign is used to describe a symbolic equation. Such an equation will evaluate to either **True** or **False**. Thus, if n is an even integer, the equation is **True** and the function returns half of n (which is still an integer since n is even). If instead, n is an odd integer, the function returns $3n+1$.

Example 5.3.2

In[283]:= `(* using If to define a function *)`
    `g[n_] := If[Mod[n, 2] == 0, n / 2, 3 n + 1]`

We can also nest **If** functions to achieve more than a "2-way" branching. Example 5.3.3 defines a function that is 1 to the right of 0, 0 at 0 and -1 to the left of 0. The first condition is $x>0$. If this is true, we return 1. But if it is false, we go to a second **If** function to further decide if x is negative or zero.

Example 5.3.3

In[284]:= `(* nesting If functions to achieve 3-`
    `way branching *)`
    `h[x_] := If[x > 0,`
      `1,`
      `If[x < 0, -1, 0]`
      `]`

Mathematica provides several other *conditionals*, that is, statements that alter the flow of execution in a program. These include **Which, Switch**, and **PieceWise**. As you learn to program more you'll want to add these functions to your repertoire, but as a beginner you should first master the **If** construction (and the use of nested

If's). You can get a lot of milage out of **If** before it is necessary to use the other conditionals. The Help Files contains a very nice tutorial about using conditionals that we'll point out in the Getting Help section.

In addition to the *If-Then-Else* construction, *loops* are essential to any programming language. A loop is a construction that allows for a set of instructions, called the *body* of the loop, to be executed over and over. There are two kinds of loops: definite and indefinite loops. Definite loops are ones where the body of the loop is executed a definite number of times that is set before the program is executed. Indefinite loops are ones where the body is repeated an indefinite number of times, depending on conditions that are not necessarily known in advance. *Mathematica* has both kinds of loops. The **Do** loop is a definite loop and the **While** loop is an indefinite loop.

Let's start with the **Do** loop. In Example 5.3.4, we use a **Do** loop to print out the first five positive integers and their cubes. The **Do** loop takes two arguments: the body, which in this case is the single line **Print**[$\{i, i^3\}$]; and the control list $\{i, 1, 5\}$, which is very much like the indexing list we would use in a **Table** function. The variable i is the index, or *counter*, that controls how many times the body of the loop is executed. In this case the loop will be repeated 5 times. The first time the body of the loop is executed the index i is 1, the next time it is 2, and so on until i reaches 5, at which point the body of the loop is executed for the last time. Thus the loop will print each of the integers and its cube from 1 to 5 and then stop. This is a definite loop. It was set up to repeat 5 times. Just as in the **Table** function, the index can be given an optional stepsize. Thus if we had used a control list of $\{i, 1, 5, 2\}$ the loop would have printed the cubes of 1, 3, and 5, skipping the cubes of 2 and 4.

Example 5.3.4

```
In[287]:= (* a definite loop that prints our the first
      five positive integers and their cubes *)
    Do[
      Print[{i, i³}],
      {i, 1, 5}
    ]
    {1, 1}
    {2, 8}
    {3, 27}
    {4, 64}
    {5, 125}
```

The body of the loop can contain any number of lines separated by semicolons. (Separating the lines by semicolons joins them together into one big expression. So **Do** still only has two arguments: the expression and then the indexing list.) Let's look at a slightly more complicated example that contains three lines of code in the body.

In Example 5.3.5, we compute the first 31 Fibonnaci numbers by using a **Do** loop. We start with the list of the first two Fibonnaci numbers, {0, 1}, and then use the loop to repeatedly append the next Fibonnaci number to the list. The body of the loop has three lines. In the first two, we let a and b be the ultimate and penultimate elements of the list so far. (On the first pass through the loop, $a=1$ and $b=0$.) Then in the third line we use the **AppendTo** function to append the next Fibonnaci number, namely the sum $a+b$, to the list. Then we repeat this for a total of 29 passes through the loop. We started with two numbers on the list and add 29 more, so we end with the first 31 Fibonnaci numbers. Compare the time spent in computing the 30th Fibonnaci number this way with the recursive definition given in Example 5.1.6.

Example 5.3.5

```
In[297]:= (* using a Do loop to compute the first 31
       Fibonnaci numbers *)
     Timing[
      fibonnaciNumbers = {0, 1};
      Do[
       a = fibonnaciNumbers[[-1]];
       b = fibonnaciNumbers[[-2]];
       AppendTo[fibonnaciNumbers, a + b]
       , {i, 1, 29}
      ]
     ]
     fibonnaciNumbers

Out[297]= {0.000373, Null}

Out[298]= {0, 1, 1, 2, 3, 5, 8, 13, 21, 34, 55, 89,
       144, 233, 377, 610, 987, 1597, 2584, 4181,
       6765, 10946, 17711, 28657, 46368, 75025,
       121393, 196418, 317811, 514229, 832040}
```

The second kind of loop is the **While** loop, which has syntax **While[**_test, body_**]**. When the loop is executed, the _test_ is evaluated. If it is **True**, the body of the loop is executed. The process is then repeated until _test_ is **False**. If _test_ never becomes

False the loop will go on forever, what we call an *infinite loop*. This is an indefinite loop because the number of passes through the loop is not specified in advance. Instead, it depends on conditions that (usually) change as the program runs.

In Example 5.3.6, we use a while loop to repeatedly apply the function g defined in Example 5.3.2 until we reach 1. In this example, we start with $k=5$. Since 5 is more than 1, we execute the body of the loop, namely, we rename k to be $3 \cdot 5 + 1 = 16$ (since 5 is odd) and print out k. Since k is still bigger than 1 we continue, this time getting $k=8$. Now 8 leads to 4 which leads to 2 which leads to 1. When k is 1, the test is no longer **True** and the loop is no longer executed.

Example 5.3.6

```
In[20]:= (* using a while loop to apply g until we
        reach 1 *)
      k = 5
      While[
        k > 1,
        k = g[k];
        Print[k]
      ]
Out[20]= 5

      16

      8

      4

      2

      1
```

In this case, starting with 5, we eventually reached 1. If we hadn't, the loop would still be running! You should rerun this example with different initial values for k. (Try $k=97$.) It turns out that no one has ever found a (positive) starting integer that did not eventually lead to 1! In fact, this is a pretty famous problem in mathematics known as the Collatz Conjecture, named after Lothar Collatz who proposed it in 1937. To date, all numbers up to some ridiculously large value have been tested. The conjecture is that no matter what positive integer we start with, repeatedly applying the function g will eventually lead to 1. It is probably true, but no one has found a proof of this!

The **If, Do** and **While** functions are tremendously important. While there are lots of other functions that are quite useful in programming, these three alone allow

Mathematica programs to be written that can carry out virtually any algorithm. We'll be seeing lot of examples in this book of their use.

5.4 Using Modules

Quite often the functions we create for ourselves can be defined in a single line of code, much like the mortgage payment function in Example 5.1.3. But what if it will take many steps to define a function? There are two possible ways to proceed.

Let's start with something simple. Suppose we are given the radius and height of a right circular cylinder and want to define a function that will return its surface area. We need to add the area of the circular top and bottom as well as the area of the sides. Example 5.4.1 shows how to define the function.

Example 5.4.1

```
In[145]:= (* this function computes the area of a right
    circular cylinder of radius r and height h *)
  area[r_, h_] := (
    baseArea = Pi r^2;
    lateralArea = 2 Pi r h;
    2 baseArea + lateralArea
  )
```

Of course, it would be really easy to define this function in one line. (We ask you how to do it in the Quiz!) But Example 5.4.1 shows how to do it in steps. We simply separate all the steps with semicolons and contain them all in parenthesis. Notice that the last "step" is simply the answer. Keep in mind that if we entered the three steps into an input cell, the last line would be evaluated and printed in the output cell. Since defining a function is really just providing a replacement rule, the answer produced by the last line will replace the function, which is exactly what we want.

Let's look at a problem where perhaps it is not so easy to define the function in a single line. Suppose we are given an integer n and want to find the nearest prime numbers to n: the largest prime less than or equal to n and the smallest prime greater than or equal to n. To find the larger prime we can start at n and, if it is not prime, start going up one number at a time until we hit a prime. Similarly, we can start at n, and if it is not prime, start going down one number at a time until we hit a prime. The definition is in Example 5.4.2.

Example 5.4.2

In[154]:= (* finds the nearest primes to n *)
```
nearPrimes[n_] := (
  k = n;
  While[PrimeQ[k] == False, k++];
  bigger = k;
  k = n;
  While[PrimeQ[k] == False, k--];
  smaller = k;
  {smaller, bigger}
)
```

In[155]:= **nearPrimes[20 123 456]**

Out[155]= {20 123 443, 20 123 479}

We begin by letting k be n. We then use a **While** loop to increase k by one at a time until we get to a prime. To do this we use the function **PrimeQ[k]** which returns **True** if k is prime and **False** if k is not prime.[3] The test condition for the **While** loop is **PrimeQ[k]==False**. Thus if k is not prime the test condition is true and we execute the body of the loop. In this case, the body is the single instruction, **k++** which increments k by 1. Thus the loop will keep going as long as k is not prime and stop when k is prime. If the original number n is prime, the test condition will be **True** right away and the body of the loop will never be executed. After the loop, we save the value of k as *bigger*. We then repeat the construction to find the smaller prime, but this time decrementing k. Finally, the last line of the definition is the output from the function, namely, the list of the smaller and bigger primes.

In both of the above examples we have introduced variables that are used as intermediate steps in order to compute the value of the function. In Example 5.4.1, these were *baseArea* and *lateralArea*. In Example 5.4.2, these were k, *smaller*, and *bigger*. These variables are only used in the definition of the function so we should not give them names that have already been used elsewhere in our notebook. If we did, we might inadvertently alter the value of a variable that is being used somewhere else. A nice way to handle this situation is to switch to defining the function as a *module*. Think of a module as a more self-contained definition of a

---

[3]There are a number of *Mathematica* functions similar to **PrimeQ** which test to see if the argument belongs in a certain set (even numbers, odd numbers, prime numbers, and the like). These functions all end in **Q**. Try entering **?*Q** to find out what they all are.

function, a kind of "black box" whose inner workings are totally separate from the rest of the notebook. Everything that takes place inside the module is completely invisible to the rest of the notebook.

In Example 5.4.3, we rewrite the **nearPrimes** function using the **Module** function. We also simplify things by eliminating *k*. The structure is nearly the same as before except that the first argument of **Module** is a list of the *local variables*, that is, those variables that are used only inside the module to help carry out the calculation. Thus the first argument in our example is {**bigger, smaller**}. A comma then separates the first argument from the second argument, which in this cases consists of five lines separated by semicolons. When expressions are separated by semicolons they are treated as one big expression. **Module** always takes two arguments: the list of local variables and then a single expression. In practice, this single expression is almost always many lines of code separated by semicolons. The last line is almost always the output of the function defined by the module, but it is possible for the output to not be on the last line.

Example 5.4.3

```
(* the nearPrimes function rewritten as a module *)
nearPrimes[n_] := Module[
  {bigger, smaller},
  bigger = n;
  While[PrimeQ[bigger] == False, bigger++];
  smaller = n;
  While[PrimeQ[smaller] == False, smaller--];
  {smaller, bigger}
]
```

Notice that the syntax coloring for Example 5.4.2 is different than Example 5.4.3. In the module example the local variables remain colored gray, further emphasizing their distinction from variables elsewhere in the notebook. Because local variables inside a module are kept separate from variables outside the module, it is perfectly okay to use variable names that might also be used outside the module. They may have the same name, but they are separate variables.

As a final example, let's rewrite the Sieve of Eratosthenes given in the Quiz in Chap. 3 as a function using **Module**. The code is in Example 5.4.4. Let's go through the example and see how it works. We name the function **sieve** and it has one argument *n*. The function is going to return a list of all primes up to *n*. We use

Module to define the function and the first argument is the list of local variables. In this case there are only two, **net** and **j**.

Example 5.4.4

```
In[83]:= (* Sieve of Eratosthenes *)
   sieve[n_] := Module[
     {net, j},
     (* initialize the sieve *)
     net = Range[n];
     (* for each k from 2 to n/2 see if it is
       "crossed out." If not,
     cross out all of its multiples *)
     Do[
      If[
        (* see if k has been crossed out *)
        net[[k]] ≠ 0,
        (* cross out all its multiples *)
        j = 2 * k;
        While[j ≤ n, net[[j]] = 0; j += k]
       ]
       , {k, 2, n / 2}
     ];
     (* sort and drop 0 and 1 *)
     Drop[Union[net], 2]
    ]
```

The first line initializes *net* by setting it equal to the list of consecutive integers from 1 to *n*. We use the **Range** function to do this. The next instruction is a **Do** loop. The loop is indexed by *k* which advances from 2 to *n/2*. When *k* is 2, we are going to "strike out" all multiples of 2, except 2 itself. Then when *k* is 3, we are going to strike out all multiples of 3, and so on. After we are done striking out all the multiples of a number, we advance to the first number that has not yet been struck out. That number must be prime since it is not a multiple of any smaller number. (If it were, it would have been struck out.) We'll keep it, but strike out all of its multiples. When we are done, the numbers that have not been struck out are the primes.

As the process proceeds, we will strike out a number from the list by replacing it with a zero. Thus our first line in the body of the **Do** loop begins **If[net[[k]]≠0, ...** So, if the condition is **True**, that means that *k* is prime and we need to strike out all of its multiples. If it is **False**, then k has been struck out and we can go onto the next number. The **If** function makes up the entire body of the **Do** loop. If *net[[k]]* is in fact zero, then nothing happens on this pass through the loop and we go onto the next value of k. This continues until k is more than $n/2$. We can stop at this point since the first multiple of a number more than $n/2$ is already bigger than n and so out of the range of the sieve. All that remains is to see what is done if in fact k is prime. In this case, we execute the **While** loop. Before entering the loop we set $j=2$. While the multiple $j\,k$ of k is still within the range of the sieve, we strike it out and go onto the next multiple by incrementing j. The construction $j^+ = k$ will increment j by k. Using $j^+ = 1$ is equivalent to j^{++}.

When the **Do** loop is over, *net* will contain a zero in the position of every composite number. Only the prime numbers will be left. (Actually, 1 will also be left.) The function **Union[net]** will now return the list sorted and free of duplicates. Thus it will begin $\{0,\ 1,\ 2,\ 3,\ 5,\ 7,\ \ldots\}$. Using **Drop** to eliminate the zero and the one in the first two positions, we are left with all the primes less than or equal to n.

As a final note on Example 5.4.4, notice that the index k used in the **Do** loop is not given in the list of local variables used by the module. Counting indices of this kind are always local to the function in which they are used, so do not need to be explicitly declared as local variables.

Before ending the chapter, let's do one more example. Given a starting integer k, the Collatz Conjecture says that the sequence of numbers generated by repeatedly applying the function g defined in Example 5.3.2 will lead to 1. This sequence is called the *orbit* of k. We'll talk more about orbits in Chap. 10.

Example 5.4.5

```
In[54]:= (* computing the Collatz orbit of k *)
    k = 27;
    orbit = {k};
    While[k > 1, AppendTo[orbit, k = g[k]]]
    orbit
    ListPlot[orbit,
     Joined → True,
     PlotRange → {
       {0, Length[orbit]}, {0, Max[orbit]}
      }
    ]
```

Example 5.4.5 (Continued)

Out[57]= {27, 82, 41, 124, 62, 31, 94, 47, 142, 71, 214,
107, 322, 161, 484, 242, 121, 364, 182, 91,
274, 137, 412, 206, 103, 310, 155, 466, 233,
700, 350, 175, 526, 263, 790, 395, 1186, 593,
1780, 890, 445, 1336, 668, 334, 167, 502, 251,
754, 377, 1132, 566, 283, 850, 425, 1276, 638,
319, 958, 479, 1438, 719, 2158, 1079, 3238,
1619, 4858, 2429, 7288, 3644, 1822, 911, 2734,
1367, 4102, 2051, 6154, 3077, 9232, 4616, 2308,
1154, 577, 1732, 866, 433, 1300, 650, 325, 976,
488, 244, 122, 61, 184, 92, 46, 23, 70, 35,
106, 53, 160, 80, 40, 20, 10, 5, 16, 8, 4, 2, 1}

Example 5.4.5, which is a slight modification of Example 5.3.6, will save the orbit so that we can plot it using **ListPlot**. The example follows the orbit of 27. In the first two lines we set the value of k and initialize *orbit* to be $\{k\}$. We then use a **While** loop to repeatedly apply g so long as $k>1$. We could have used **While[k>1, k=g[k]]** to do this, but then there would be no record of the orbit. So, during each pass through the loop we use **AppendTo** to append the next step in the orbit to the list *orbit*. Notice that we both name k to be the next element of the orbit as well as append it to the orbit list in a single step. We could just as well have written **While[k>1, k=g[k]; AppendTo[orbit, k]]**.

The next line, **orbit**, will cause the orbit to be printed out. We end with **ListPlot** which plots the orbit. So far we have used **ListPlot** by passing in a list of points to be plotted. Here we have only passed in a list of numbers. If you check the syntax of **ListPlot** you'll see that in such a case *Mathematica* will plot points having these numbers as their *y*-coordinates, with the *x*-coordinate for each number being its place in the list. We use the option **Joined** → **True** to connect the points with line segments. Finally, notice the use of **PlotRange** and the upper limits in both the horizontal and vertical direction that will change automatically and remain correct as we experiment with different values for *k*.

From the plot, we can see the wild ride taken by the orbit of 27 as it climbs and falls, eventually ending at 1.

In Example 5.4.6, we use **Module** to define the function **collatzStoppingTime[*n*]** that will compute the length of the orbit of *n*.

Example 5.4.6

```
In[43]:= (* this function will give the minimum
          number of iterations to reach 1 starting
          from n *)
       collatzStoppingTime[n_] := Module[
         {k = n, count = 0},
         While[k > 1, k = g[k]; count++];
         count
         ]

In[59]:= collatzStoppingTime[27]

Out[59]= 111
```

The heart of the function is the **While** loop wherein we follow the orbit until we reach 1. This time however, we want to count how many passes through the loop are taken. We create a local variable *count*, set it to 0 before we enter the loop, increment it each time we pass through the loop, and return it on the last line of the **Module**. A nice option when using **Module** is that we can set the initial values of local variables as we define them in the first argument of **Module**. So, we could have begun the module as

```
collatzStoppingTime[n_] := Module[
  {k, count},
  k = n;
  count = 0;
```

but it is simpler to set the initial values of *k* and *count* as we list them in the list of local variables. Notice that we could use this simplification in Example 5.4.3, setting both *bigger* and *smaller* equal to *n* when they are introduced in the list of local variables. We could then eliminate two lines of code from the body of the loop. Similarly, in Example 5.4.4 we could have initialized *net* in the list of local variables. The local variable list would then appear as {net = **Range**[n], j}.

   Notice the use of *k* to take the place of *n* right from the begining. It will not work if we try

```
In[61]:= collatzStoppingTime[n_] := Module[
       {count = 0},
       While[n > 1,  n = g[n]; count ++];
       count
       ]
```

Try it! The problem is with *n=g[n]*. Since *n* is the input to the function we can use it all we want on the right-hand sides of equals signs, but we cannot reset the value of the input to the function.

   In Chap. 10, we'll introduce the function **NestList**. This will give a much better way of following orbits than Examples 5.4.5 or 5.4.6.

5.5 Find Out More

As always, there are lots of pages in the Help Files worth looking at. Here are a few suggestions.

- tutorial/PatternsOverview—excellent discussion of patterns and their use in defining functions.
- tutorial/FunctionsAndProgramsOverview—a very nice tutorial on defining functions.
- tutorial/SettingUpFunctionsWithOptionalArguments—just what it says. It's part of the larger tutorial/PatternsOverview.
- guide/LayoutAndTables—a nice guide on laying out tables.
- tutorial/Conditionals—part of the larger tutorial/EvaluationOfExpressions-Overview, this is a nice tutorial on conditionals, statements that allow branching in programming.
- tutorial/LoopsAndControlStructures—an excellent discussion of loops. It's part of the larger tutorial/EvaluationOfExpressionsOverview.

   In addition to the above, there is a nice article in *Wikipedia* on the Collatz Conjecture.

Quiz

1. Define a function (in a single line) that will compute the surface area of a right circular cylinder of radius r and height h.

2. After using the mortgage **payment** function defined in Example 5.1.3, you might decide that the number of months to pay off the loan (currently set to 360) should also be an argument to the function. Rewrite the function to accept three arguments: the principal P, interest rate r, and the number of months to pay off the loan, m.

3. Define a function that will take a year such as 2008 and return **True** if it is a leap year and **False** if it is not. (You'll need to look up the exact definition of leap years to do this!)

4. Define the function **dayCount[day_, month_, year_]**, that will return the number of days from the beginning of that year to, and including, the given date. For example, **dayCount[6, 1, 2007]** is 6.

5. Suppose $data = \{\{r_1, \theta_1\}, \{r_2, \theta_2\}, \dots, \{r_n, \theta_n\}\}$. Write a pure function **expr&** so that **Map[expr&, data]** produces the list $\{r_1 \sin\theta_1, r_2 \sin\theta_2, \dots, r_n \sin\theta_n\}$.

6. Create an input cell whose output cell is the multiplication table for integers 1 through n for any n. The i, j entry in the table should be the product $i \cdot j$ and the rows and columns should be labeled with the factors i and j.

7. Modify **nearPrimes[n]** in Example 5.4.3 to return the distance from n to the nearest prime.

8. Create a function that will take two positive integers n and m as input and return the list of primes p such that $n \le p \le m$. (Hint: Use **PrimeQ**.)

9. Use a **While** loop to find the smallest integer n such that **collatzStoppingTime[n]** is equal to 200. (Hint: Start at 1 and keep going up until you reach an integer that has the desired property.)

10. The ancient Babylonians devised the following algorithm for finding the square root of a. First start with a reasonable guess and call it x_0. If x_0 really were the square root of a, then $x_0^2 = a$, or equivalently, $x_0 = \frac{a}{x_0}$. But since our guess is not likely to be the square root of a, the numbers x_0 and $\frac{x_0}{a}$ will be different. Hence, let's use their average as our next guess. This gives

$$x_1 = \frac{x_0 + \frac{a}{x_0}}{2}$$

We then repeat the process using x_1 to produce x_2, and so on, with each successive step taking us closer to the square root of a.

For example, suppose we want to find the square root of 5. If we start with a guess of $x_0 = 2$, this leads to $x_1 = (2 + 5/2)/2 = 9/4$, which leads to $x_2 = (9/4 + 5/(9/4))/2 = (9/4 + 20/9)/2 = 161/72$, and so on. This is already quite close to the square root as $(161/72)^2 - 5 = 1/5184$.

Define a function that will use the Babylonian algorithm to find the square root of any number. In particular, after entering a and x_0 into the function, continue to calculate successive approximations $x_0, x_1, x_2, \ldots$ until x_i^2 is within 10^{-5} of a.

CHAPTER 6

Three-Dimensional Graphics

In Chap. 2, we saw how to plot a variety of two-dimensional objects. Almost all of the functions that were discussed there have a version that works similarly for three-dimensional objects. In this chapter we'll see how to plot curves and surfaces in three-dimensional space as well as other kinds of objects.

6.1 The Plot3D Function

Let's start with the **Plot3D** function which is the analog of the **Plot** function. We can use **Plot3D** to plot the graph of a two-variable function in three-dimensional space. Example 6.1.1 shows the graph of the function $x^2 + y^2$. The syntax for **Plot3D** is very similar to that of its two-dimensional cousin **Plot**. The first argument is the function that we want to plot the graph of, in this case $x^2 + y^2$. But now there are *two* independent variables, x and y, so the next thing we must tell *Mathematica* is the domain of *each* variable. We do this with the two lists {**x**, -2, **2**} and {**y**, -2, **2**}. *Mathematica* draws a picture of the graph in perfect perspective as seen from a point called the **ViewPoint**. (Check out **ViewPoint** in the Help Files. You'll find that you can easily change this parameter, and hence the view of the object.) It also shades the surface and hides parts from view should one part of the surface be in front of another part.

One of the really neat things about *Mathematica* is that after the graph of the surface is drawn we can grab it with the mouse and spin it around! Try moving the mouse over the graphic. The cursor will turn into a pair of curved arrows each pointing to the tail of the other. If you hold down the mouse button and drag the graphic you can rotate the surface in space! If you additionally hold down the Option or Command key you can zoom in or out. Holding down the Shift key while dragging allows you to drag the graph without rotating it.

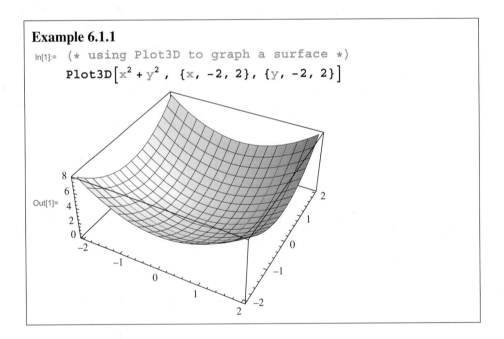

Example 6.1.1

In[1]:= (* using Plot3D to graph a surface *)
        Plot3D$\left[x^2 + y^2 , \{x, -2, 2\}, \{y, -2, 2\}\right]$

Just like **Plot**, **Plot3D** will allow us to plot the graph of more than one surface at once. We simply replace the function we want to plot with a list of functions. Example 6.1.2 shows the graphs of both $x^2 + y^2$ and $4 - x^2$ simultaneously.

Example 6.1.2

In[2]:= (* graphing multiple functions *)
   Plot3D[
      {x² + y² , 4 - x²}, {x, -2, 2}, {y, -2, 2}
   ]

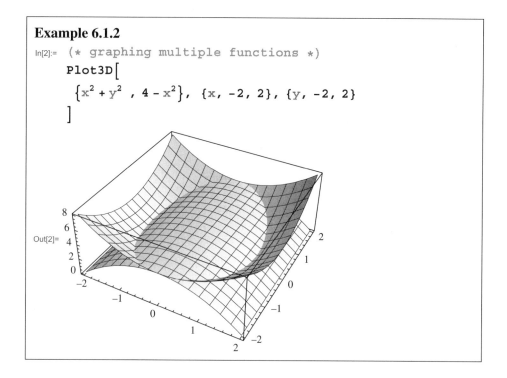

Out[2]=

6.2 3D Graphics Options

We saw in Chap. 2 that we could change the appearance of a graph in all sorts of ways, either by using plot options or by using the 2D Drawing Tools or 2D Graphics Inspector. We can change a 3D plot too, but only by using options, there is no version of the Drawing Tools or Graphics Inspector for 3D. So, we see another reason to learn to use both plot options as well as the Drawing Tools.

Example 6.2.1

In[3]:= (* using BoxRatios to control aspect
      ratios *)
   Plot3D[
      {x² + y² , 4 - x²}, {x, -2, 2}, {y, -2, 2},
      BoxRatios → {1, 1, 2}
   ]

Example 6.2.1 (Continued)

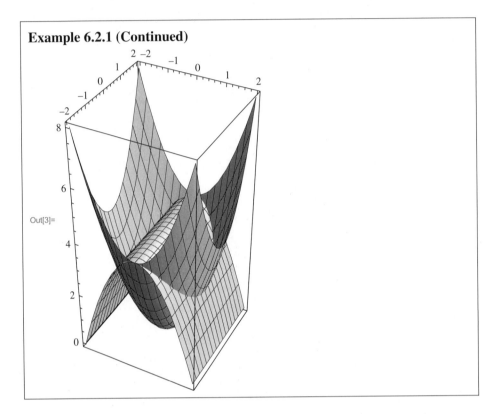

Out[3]=

    Notice that in both Examples 6.1.1 and 6.1.2 the scale in the vertical direction is
not the same as in the two horizontal directions. This was often true of 2D plots too
and in that case we could adjust the scales by using the **AspectRatio** option. Since
our plot now represents a three-dimensional object, surrounded by a *bounding box*,
there are really three separate aspect ratios possible, namely, height to width, height
to length, or length to width. To control these we can use the **BoxRatios** option.
Suppose we want to make the scales on the three axis in Example 6.1.2 all appear
to be the same length. We can either use the option **BoxRatios→Automatic** or the
more direct **BoxRatios→{1, 1, 2}** where we specifically tell *Mathematica* to make
the bounding box twice as high as it is wide or long. (Since the vertical direction
in the example spans 8 units, from 0 to 8, while each of the horizontal directions
spans 4 units, from -2 to 2, we need the bounding box to be twice as tall as it
is wide.) The first way is nice because it will keep the size of the bounding box
correctly chosen if we should later change the domain of x and y. Example 6.2.1
shows the result.
    The surface which we graphed in Example 6.1.1 is called a *paraboloid*. It can
be obtained as a surface of revolution by rotating a parabola around its axis of
symmetry. We can emphasize this property of the surface if we graph it over a disk

rather than a square in the xy-plane. To do this we can use the **RegionFunction** option. The **RegionFunction** option will allow us to plot the graph of a function over any region we specify in the xy-plane. For example, suppose we want to graph $x^2 + y^2$ over the disk of radius 2 centered at the origin. The following example shows how to do this by using the option **RegionFunction→Function[{x, y, z}, x²+y² <4]**. To specify a different region we would just change the part where we have written **x²+y² <4** and keep everything else the same.[1]

Example 6.2.2

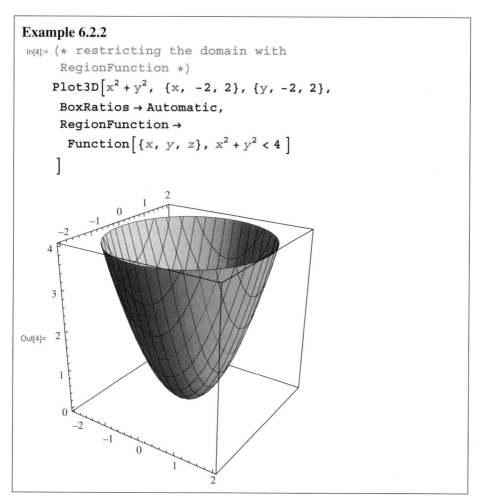

```
In[4]:= (* restricting the domain with
          RegionFunction *)
        Plot3D[x² + y², {x, -2, 2}, {y, -2, 2},
          BoxRatios → Automatic,
          RegionFunction →
            Function[{x, y, z}, x² + y² < 4]
        ]
```

Out[4]=

---

[1]Remember that the circle of radius r centered at the point (h, k) has equation $(x - h)^2 + (y - k)^2 = r^2$. Thus the circle with radius $r = 2$ and center $(h, k) = (0, 0)$ has equation $x^2 + y^2 = 4$. The closed *disk* of all points *inside* the circle together with the circle itself would be given by the inequality $x^2 + y^2 \leq 4$. If we replaced the less than or equal sign, $\leq$, with a strictly less than sign, $<$, we would then have the *open* disk consisting of only the points inside the circle.

There are more than 20 options that can be used with **Plot3D**. Evaluate **Options[Plot3D]** to see what they all are, or take a look at the Help Files under both **Plot3D** and **Graphics3D**. We'll close this section with two more examples, that illustrate some of the options.

In Example 6.2.3, we have modified the **RegionFunction** option in an interesting way and have also introduced the **PlotStyle**, **BoundaryStyle**, and **Mesh** options. Starting with **RegionFunction**, we have replaced $x^2 + y^2 < 4$ with $x^2 + y^2 < 4$ && $(x > 0 || y > 0)$ && $(1 < z < 2 || 3 < z < 4)$. This describes the points to include

---

Example 6.2.3

```
In[34]:= (* using PlotStyle, BoundaryStyle, & Mesh *)
    Plot3D[x² + y², {x, -2, 2}, {y, -2, 2},
     BoxRatios → Automatic,
     RegionFunction → Function[
        {x, y, z}, x² + y² < 4 &&
        (x > 0 || y > 0) &&
        (1 < z < 2 || 3 < z < 4)
      ],
     PlotStyle → {Opacity[.5], Blue},
     BoundaryStyle → {Red, Thickness[.01]},
     Mesh → None
    ]
```

Out[34]=

in the plot by using three different conditions combined with the "logical and" represented by the double ampersands, **&&**. *Mathematica* will only plot points where all three conditions are true. The first condition is the same as what we had before, namely, $x^2 + y^2 < 4$. But we have added to this the second condition (**x>0 || y>0**). This condition also uses a logical operator, the "logical or," ||. Thus the second condition asserts that either $x > 0$ or $y > 0$ (or both are greater than zero). This rules out both x and y being negative or zero and hence we do not see any surface over the "third quadrant" of the xy-plane. Finally, the third condition says that z must either be between 1 and 2, or be between 3 and 4.[2]

The **PlotStyle** option allows us to alter the plot in many ways. In this case, we have chosen to change the opacity and the color of the plot. The opacity of the surface is a measure of its transparency, with **Opacity[0]** being completely transparent and **Opacity[1]** be opaque. We have also told *Mathematica* to color the surface blue by using the color name **Blue** (instead of **RGBColor[0, 0, 1]**). *Mathematica* has conveniently provided a handful of named colors:

Red	Green	Blue	Black
White	Gray	Cyan	Magenta
Yellow	Brown	Orange	Pink
Purple	LightRed	LightGreen	LightBlue
LightGray	LightCyan	LightMagenta	LightYellow
LightBrown	LightOrange	LightPink	LightPurple

Each of these has a corresponding **RGBColor** code, or in the case of Black, White, Gray, and LightGray, a **GrayLevel** code. If you want to find out what the code is just enter a color name into a cell and evaluate it. The output will be the corresponding color code.

Additionally, we have used **BoundaryStyle** to change the appearance of the edge, or boundary, of the surface. In this case, we have colored it **Red** and given it a **Thickness** of .01. Finally, **Mesh→None** has turned off the gridlines, or mesh, that is normally drawn on the surface. We can not only turn the mesh on or off, but also space the mesh lines as we please or color the surface checkerboard fashion between the mesh lines.

We close this section with one more example that illustrates some neat effects that can be gotten using plot options. In Example 6.2.4, we have switched to a new function, x^2, as well as thrown in some new plot options. **Filling** can be used to fill

---

[2]If P and Q are "statements," then the *compound statement* $P \&\& Q$ (which we read "P and Q") is true exactly when **both** P and Q are true. The compound statement $P||Q$ (which we read "P or Q") is true exactly then **either** P or Q or **both** are true.

in the space between the surface and some other place. In this case by specifying **Filling** → **Bottom** *Mathematica* has filled in the space between the surface and the bottom of the bounding box. We have also used **FillingStyle** so that the filling is **Green** and has **Opacity[.3]**. Since the filling is nearly transparent it may be difficult to see in the figure. (**Filling** can also be used in two-dimensional plots with **Plot**. Try it! Look up **Filling** in the Help Files to see what values it can take.) You should try this example and play with the opacity parameter. Interestingly, with an opacity of zero (transparent) the green disk at the bottom of the bounding box is still shown. So this is a neat way to color in the domain of the function in the xy-plane.

Example 6.2.4

```
In[1]:= (* Filling in below the surface *)
    Plot3D[x^2, {x, -2, 2}, {y, -2, 2},
      BoxRatios → Automatic,
      RegionFunction →
        Function[{x, y, z}, x^2 + y^2 < 4],
      BoundaryStyle → {Red, Thickness[.01]},
      Filling → Bottom,
      FillingStyle → {Green, Opacity[.3]},
      Mesh → 8,
      MeshShading → {{Black, None}, {None, Black}}
    ]
```

We have also used the **Mesh** option, together with **MeshShading**, to color the surface checkerboard style. The option **Mesh→8** tells *Mathematica* to place 8 mesh lines in each direction. The option **MeshShading→{Black, None},{None, Black}}** then causes the spaces between the mesh lines to alternately be colored **Black** and **None** in each of the x and y directions.

There are simply too many plot options to explain them all here. You should look at the Help Files which are filled with lots and lots of examples. The main thing to keep in mind is: If you can imagine it, *Mathematica* can probably do it.

6.3 Surfaces of Revolution

As we already mentioned, the paraboloid of Example 6.1.1 is a surface of revolution obtained by revolving the graph of x^2 around its axis of symmetry. *Mathematica* has a built in function, **RevolutionPlot3D** that will plot surfaces of revolution. In Example 6.3.1 we revolve the graph of $\cos x$ around the z-axis.

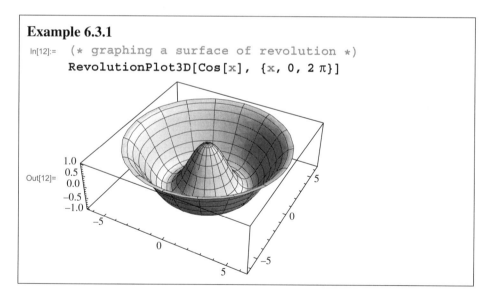

Example 6.3.1

In[12]:= (* graphing a surface of revolution *)
          RevolutionPlot3D[Cos[x], {x, 0, 2 π}]

Out[12]=

In this example the arguments of the **RevolutionPlot3D** function are the *generating curve* **Cos[x]** and its domain list {**x, 0, 2π**}, exactly the same arguments we would use in the **Plot** function if we wanted to plot the curve $z = \cos x$ in the xz-plane. However, instead of plotting this curve, **RevolutionPlot3D** spins the curve around the z-axis and plots the resulting surface. If we wanted to revolve a different curve around the z-axis we would replace **Cos[x]** with the new equation of the generating curve. Of course, all the usual plot options can be used with

RevolutionPlot3D even though in this example we did not use any. It is also worth mentioning that **RevolutionPlot3D** can only be used to spin around the z-axis.

In addition to the list $\{x, 0, 2\pi\}$, which defines the domain of the generating curve, we may also enter, if we want, the *angular domain* of the plot. For example, suppose we only want to spin the generating curve half way around the z-axis. We simply follow the curves domain with a similar list for the domain of the angle. The actual variable that we use for the angle can be anything. Example 6.3.2 illustrates this as well as two useful options for turning off the bounding box and the axes.

Example 6.3.2

```
In[2]:= (* restricting the angular domain *)
     RevolutionPlot3D[
      Cos[x], {x, 0, 2 π}, {θ, 0, π},
      Boxed → False,
      Axes → False
     ]
```

Out[2]=

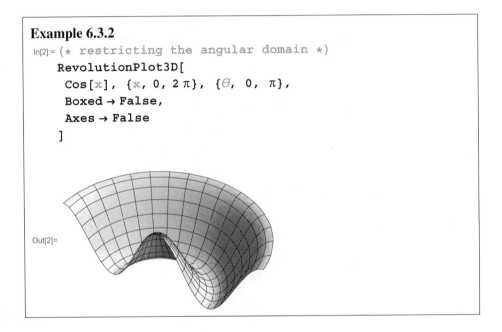

The option **Boxed** $\rightarrow$ **False** causes the bounding box to not be drawn (as opposed to the default **Boxed** $\rightarrow$ **True**) and the option **Axes** $\rightarrow$ **False** causes the axes, with tick marks and scales, to not be drawn (as opposed to the default **Axes** $\rightarrow$ **True**).

So far we have seen how to use **RevolutionPlot3D** by describing the generating curve as the graph of a function of one variable. But what if the generating curve is not the graph of a function? For example, suppose we want to revolve a circle around a line to generate a *torus*, or doughnut-shaped surface. We cannot describe the circle as the graph of a function, but we can describe it parametrically, and fortunately for us, **RevolutionPlot3D** will accept a description of the generating curve parametrically too. Consider the circle in the xz-plane centered at $(2, 0)$ and with radius 1. We can describe its (x, z)-coordinates parametrically as $(2 + \cos t, \sin t)$, where the parameter t goes from 0 to 2π. How to revolve this curve around the z-axis is illustrated in Example 6.3.3.

Example 6.3.3

In[3]:= (* describing the generating curve
      parametrically *)
    RevolutionPlot3D[
      {2 + Cos[t], Sin[t]}, {t, 0, 2 π}
    ]

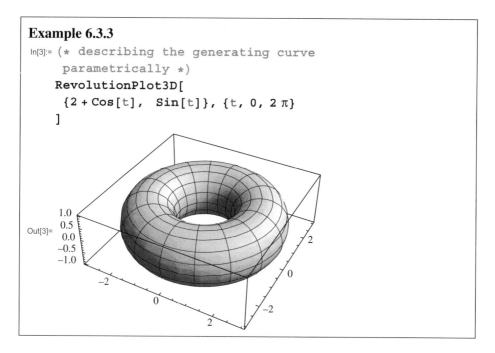

Notice that the first argument of **RevolutionPlot3D** is a list of functions, and with the functions **Plot** or **Plot3D** we use a list of functions as the first argument when we want to plot more than one curve or surface simultaneously. So the syntax of **RevolutionPlot3D** is slightly different, and the reason is so that we can give the generating curve parametically if we want to. It is possible to plot more than one surface of revolution simultaneously and to do so we provide a list of each generating curve. But now each generating curve must itself be contained in a list. Thus to spin both $z = x^2$ and $z = \cos x$ around the z-axis simultaneously we would write

$$\textbf{RevolutionPlot3D}[\{\{\textbf{x}^2\},\{\textbf{Cos}[\textbf{x}]\}\}, \{\textbf{x, 0, 2}\pi\}]$$

If we want to spin more than one generating curve, and describe one of them parametrically, then we need to describe all of them parametrically (and have the same parameter and domain for each). Try evaluating the following:

$$\textbf{RevolutionPlot3D[}$$
$$\{\{\textbf{t, t}\},\{\textbf{t,-2 t}\},\{\textbf{4+Cos[t], Sin[t]}\}\},$$
$$\{\textbf{t, 0, 2 Pi}\}$$
$$\textbf{]}$$

You should see two cones joined at their vertex surrounded by a torus. Each of the three generating curves is described parametrically: (t, t) describes a line that sweeps out a cone above the xy-plane; $(t, -2t)$ describes a line that sweeps out a cone below the xy-plane; $(4 + \cos t, \sin t)$ describes a circle that sweeps out a torus.

We can make Example 6.3.3 a little more interesting by adding a few options as the next example shows.

Example 6.3.4

```
In[4]:= (* using options to jazz up the plot *)
    RevolutionPlot3D[
     {2 + Cos[t],  Sin[t]}, {t, 0, 2 π},
     Boxed → False,
     Axes → False,
     Mesh → 9,
     MeshShading → {
       {GrayLevel[.9], None}, {None, GrayLevel[.9]}
      }
    ]
```

Out[4]=

The option **Mesh**→**9** caused 10 gridlines in each direction to be drawn and **MeshShading**→{{**GrayLevel[.9], None**},{{**None, GrayLevel[.9]**}}} caused the mesh to be colored checkerboard fashion with squares that alternate from clear to gray.

6.4 Drawing Contours or Level Sets

There are many different ways that we can try to visualize a function. One way, which we have already discussed in both this chapter and Chap. 2, is to draw the graph of the function. Functions like **Plot**, **ParametricPlot**, and **Plot3D** do just that.

But another way to visualize a function is to draw its *contours* or *level sets*. A single contour, or level set, is a subset of the domain of a function where the function is constant. In the case of a function of two variables a single contour is typically a curve in the domain. Weather maps that show air pressure are perhaps a familiar example. In this case, the contours are also known as "isobars," since everywhere along a single contour the pressure is equal. Topographic maps used for hiking and orienteering also provide a good example. In this case, the contour lines represent places where altitude above sea level is constant.

We can use *Mathematica* to draw contour diagrams of functions of both two and three variables by using the functions **ContourPlot** and **ContourPlot3D**. Let's start with something simple. Example 6.4.1 shows the contour lines of the function $x^2 + y^2$. We have already seen the graph of this function in Example 6.1.1. It is the surface of revolution obtained by spinning the graph of $z = x^2$ around the z-axis. Since it is a surface of revolution its contour lines must be circles!

Example 6.4.1

In[5]:= (* contours of the paraboloid are circles *)
        ContourPlot$\left[x^2 + y^2, \{x, -2, 2\}, \{y, -2, 2\}\right]$

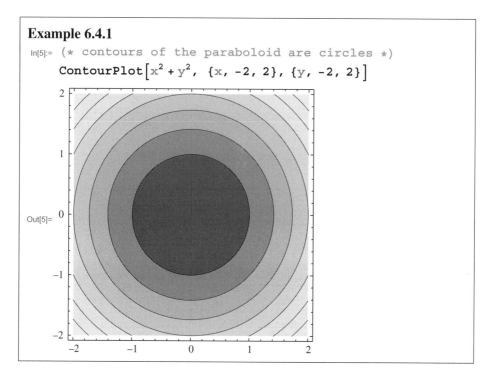

Out[5]=

In this example we see a set of concentric circles with a coloring scheme that is dark in the center and becomes progressively lighter as we move out. Each circle is a contour, or level set. The value of the function is constant along each contour. A really neat feature of **ContourPlot** is that *Mathematica* will display the value of the function at a specific contour if we move the mouse pointer to that contour. Try it! In this case, we can discover that the values of the function at each contour as we start in the center and move out are 1, 2, 3, 4, 5, 6, and 7. If you are practiced with topographic maps then it is not hard to imagine the graph of the function. It is low at the origin and then becomes higher as we move out. Since the contour lines are not equally spaced we can also tell that the graph is getting steeper as we move away from the center.

As with every graphics function we have seen so far, we can put in lots of options to customize the plot. What kinds of things would be worth changing in a contour plot? It would certainly depend on what you were making the plot for, but at least two items that might be useful to customize are the domain of the function and the values of the function where the contours are drawn.

Changing the domain is easy. Of course, when we provide the domain lists for each of the two variables we are tacitly providing a rectangular domain. In Example 6.4.1, we let both x and y run from -2 to 2. But we can use the option **RegionFunction** to further restrict the domain. Example 6.4.2 illustrates this.

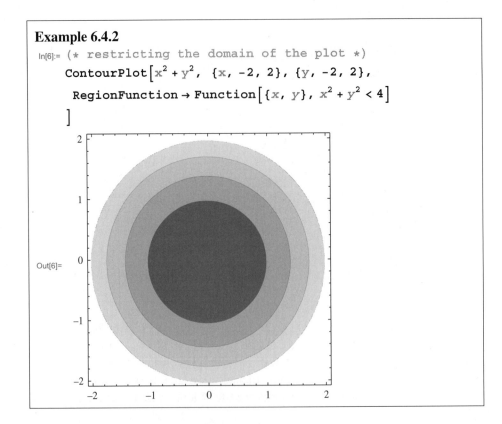

Example 6.4.2

In[6]:= (* restricting the domain of the plot *)
ContourPlot$\left[x^2 + y^2, \{x, -2, 2\}, \{y, -2, 2\}, \right.$
RegionFunction $\rightarrow$ Function$\left[\{x, y\}, x^2 + y^2 < 4\right]$
]

Out[6]=

To control where *Mathematica* places the contours we can use the **Contours** option. Example 6.4.3 shows the contours of the function $x^2 - y^2$ at every integer from -3 to 3. We have also used the option **ContourLabels→Automatic** so as to label each contour with the value of the function there. *Mathematica* decides where to place the labels so as to maximize readability. If this were a topographic map, the experienced hiker could tell that we were looking at a saddle or mountain pass. The saddle connects two valleys located at the top and bottom of the map (and colored dark). The saddle is a low point on a ridge that runs from the left to the right sides of the map.

Example 6.4.3

In[7]:= `(* labeling the contours *)`

```
ContourPlot[x² - y², {x, -3, 3}, {y, -3, 3},
 Contours → {-3, -2, -1, 0, 1, 2, 3},
 ContourLabels → Automatic
]
```

Out[7]=

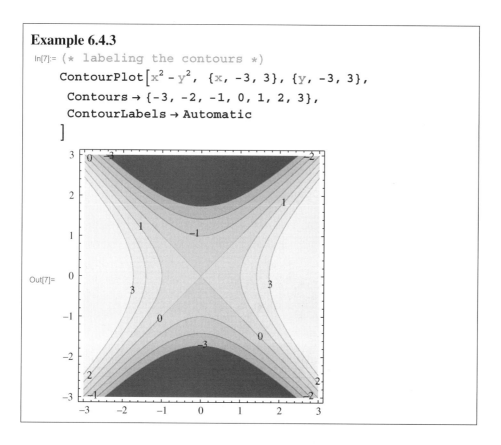

If we are given a function of two variables we now have two ways to visualize it. We can draw its graph, or we can plot its contours. Both can be quite helpful. But what if we are trying to understand a function of three variables? We can't draw its graph as that would require a fourth dimension! But we can draw its contours! In general, we should expect the contour sets to be surfaces. Again, remember that contours always lie in the domain of a function. So when we look at the contours of a three-variable function we are looking at the domain of the function with specified sets where the function is constant. Let's try it out.

Example 6.4.4 uses **ContourPlot3D** to plot the contours of the function xyz. We have also used a couple of nice options to modify the plot. We have specified that the contours should correspond to functional values of $-2, -1, 0, 1$, and 2 and have used **ContourStyle** to control the opacity of each contour surface. Finally, we have turned the mesh off with **Mesh→False**. Notice that one of the contours is the union of the three coordinate planes. In each of these planes at least one of the coordinates is zero and so the product is zero. Conversely, if $xyz = 0$ then at least one of the three variables must be zero. So the union of the three coordinate planes is the contour corresponding to a functional value of zero. But which of the contour surfaces describes where the function is 1 or 2?

Example 6.4.4

```
In[9]:= (* contours of xyz are surfaces *)
    ContourPlot3D[
     x y z, {x, -2, 2}, {y, -2, 2}, {z, -2, 2},
     Contours → {-2, -1, 0, 1, 2},
     ContourStyle → {Opacity[.8]},
     Mesh → False
    ]
```

Out[9]=

Unlike **ContourPlot**, **ContourPlot3D** does not display the value of the function as we move the mouse over the contours. Nor can we display those values with **ContourLabels** which is no longer an allowable option. We're on our own here to understand what we are looking at! Example 6.4.5 shows how to plot a single contour. Notice the use of double equals signs. Remember that a single equals sign in *Mathematica* is used for replacement, replacing the value of the variable on the left with the value of the one on the right, as in $a = 3$. Two equals signs are used when we want to test for equality. Example 6.4.5 plots the contour of xyz, where the value of the function is equal to 1.

Example 6.4.5

```
In[10]:= (* graphing a single contour *)
ContourPlot3D[
  x y z == 1, {x, -2, 2}, {y, -2, 2}, {z, -2, 2},
  ContourStyle → {Opacity[.8]},
  Mesh → False
]
```

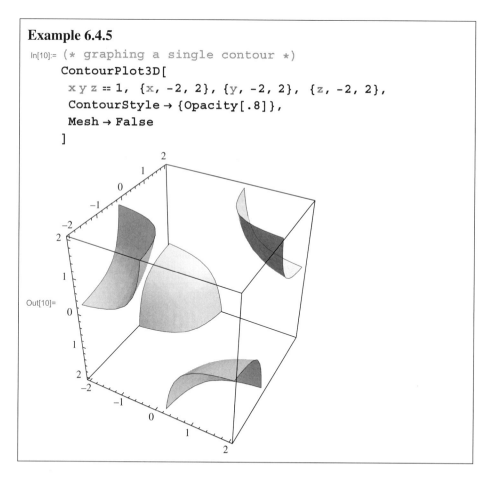

Out[10]=

If we want to see different contours corresponding to different values of the function, we can just repeat this example, replacing $xyz == 1$ with whatever value we want. What a great place to use **Manipulate**! Let's replace $xyz == 1$ with $xyz == a$ and then manipulate the parameter a. This is shown in Example 6.4.6.

Notice that instead of listing the parameter and its domain as $\{a, -1, 1\}$, we have listed it as $\{\{a, 0\}, -1, 1\}$. This form tells *Mathematica* that a should vary between -1 and 1, but that its initial value is set to zero. When the plot is rendered for the first time, it is done so with $a = 0$.

Example 6.4.6

In[11]:=
```
(* animating the contours *)
Manipulate[
 ContourPlot3D[
  x y z == a, {x, -2, 2}, {y, -2, 2}, {z, -2, 2},
  ContourStyle → {Opacity[.8]},
  Mesh → False],
 {{a, 0}, -1, 1}
 ]
```

Out[11]=

Notice that by using **ContourPlot3D**, we can graph the solution set to any equation in three variables, even if we cannot conceive of that set as the graph of a function. For example, consider the sphere with radius 4 and center at the origin given by $x^2 + y^2 + z^2 = 16$. This is not the graph of a function, so we could not plot it with **Plot3D**, for example. However, it is a single contour of the function $x^2 + y^2 + z^2$, namely, the contour where the function is equal to 16. Thus we could plot it using **ContourPlot3D** by plotting only the single contour corresponding to 16. Notice that this will work, at least in theory, for any equation in three variables, for we can always think of such an equation as representing a single contour of a function in three variables. The same goes for an equation in two variables too, as we could then use the function **ContourPlot**. For example, suppose we want to graph the solution set of the equation $xy + x^2 - \sin y = x^3$. It would not be easy to solve this for x in terms of y or for y in terms of x and hence we cannot really conceive of this as the graph of a function in one variable. But if we rewrite this as $xy + x^2 - \sin y - x^3 = 0$, then we see that this is the single contour of the two-variable function $xy + x^2 - \sin y - x^3$ corresponding to a functional value of zero.

Example 6.4.7

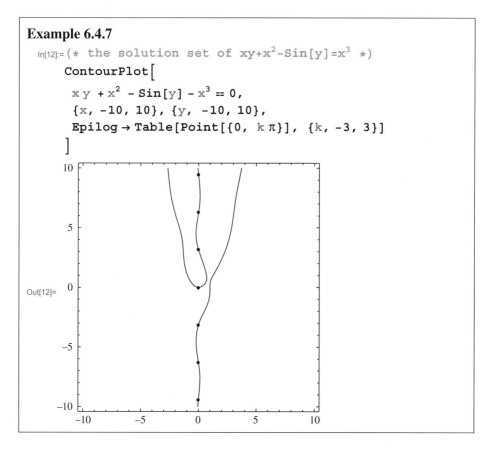

```
In[12]:= (* the solution set of xy+x²-Sin[y]=x³ *)
        ContourPlot[
          x y + x² - Sin[y] - x³ == 0,
          {x, -10, 10}, {y, -10, 10},
          Epilog → Table[Point[{0, k π}], {k, -3, 3}]
        ]
```

In Example 6.4.7, we use **ContourPlot** to graph this rather strange curve. Notice that if $x = 0$ the equation reduces to $-\sin y = 0$ and so y must be a multiple of π. So all the points $(0, 0)$, $(0, \pm\pi)$, $(0, \pm2\pi)$, and so on, must lie on the curve. In the plot we have used the **Epilog** option to plot these points after the graphic has been rendered. We use **Table** to first make a list of the points and then plot them with **Epilog**. The function **Point** is a graphics primitive.

6.5 Drawing Curves in 3-Space

Imagine a particle moving through space and tracing out a curve. Its position is a function of time and hence each of its coordinates, x, y, and z, can be described as a function of time. This gives us a parametric description of the curve with time serving at the parameter.

Example 6.5.1

```
In[13]:= (* graphing a curve in space *)
    helix = ParametricPlot3D[
        {Cos[t], Sin[t], t}, {t, 0, 2 π},
        PlotStyle → {Thickness[.02]}
    ]
```

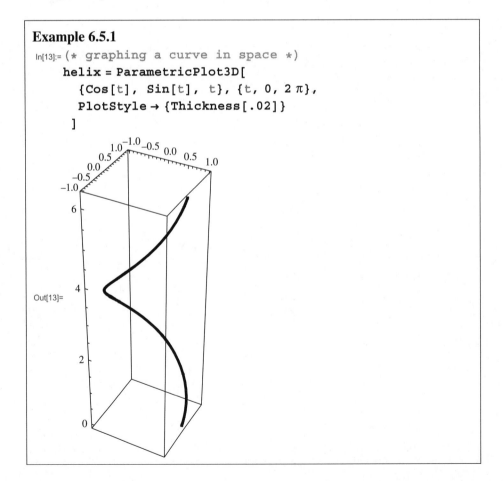

Out[13]=

For example, suppose that at time t a particle is located at the point

$$(x, y, z) = (\cos t, \sin t, t)$$

What will be the curve traced out by the particle? We can easily plot this with *Mathematica* using **ParametricPlot3D** as seen in Example 6.5.1.

This curve is called a *helix*. It winds around the z-axis like the thread on a bolt. As usual, **ParametricPlot3D** takes all the standard graphics options and we have used **PlotStyle** to thicken up the curve. In fact, the helix lies on a cylinder of radius 1 whose axis is the z-axis. Wouldn't it be cool to draw in the cylinder too, perhaps just faintly by using a low opacity? We'll do this in the next section after we see how to plot a cylinder.

Let's look at one more example of a parametric curve. A *torus knot* is a simple closed curve that lies on the surface of a torus like that pictured in Example 6.3.3. By *simple* we mean that the curve does not intersect itself, and by a *closed* curve we mean one that begins and ends as the same point. Thus a simple closed curve in space is a loop that does not intersect itself. Topologists call such a loop a *knot* and a torus knot is just a knot that lies on a torus.[3] A torus knot that winds around a torus p times in the *longitudinal* direction and q times in the *meridional* direction can be parameterized as

$$x(t) = R\cos(pt) + r\cos(pt)\cos(qt)$$
$$y(t) = R\sin(pt) + r\sin(pt)\cos(qt)$$
$$z(t) = r\sin(qt)$$

where p and q are relatively prime integers, that is, their greatest common divisor is 1. Here r is the radius of the circle used to generate the torus as a surface of revolution, and R is the distance from the center of the generating circle to the z-axis. Example 6.5.2 shows the torus knot with $p = 3$ and $q = 2$. This knot is called the *trefoil* knot. Notice that at the beginning of the cell we have given values to the variables r, R, p, and q. Then when we write the parametric formulas for the coordinates inside the **ParametricPlot3D** function we simply use expressions that contain r, R, p, and q. If we want to change any of the parameters for the knot we just need to change the values at the beginning and *Mathematica* does the rest. This is a good style that you should work to develop. Try this example out and experiment with the values of the parameters!

---

[3] *Topology* is the branch of mathematics that deals with properties of space and especially how one space can sit inside another. *Knot Theory* is the study of how a loop can be placed inside three-dimensional space.

We have used the **PlotStyle** option **Tube[.01]** which is similar to **Thickness** but which draws a tube centered on the curve. A nice feature of **Tube** is that is allows us to clearly see where the knot crosses over itself. We have also used **PlotRegion** to specifically make the bounding box large enough to contain the knot and the tube. Notice the constant of 1.1 that is multiplied times the **PlotRegion** dimensions. If this constant is too small the tube will be clipped off at the faces of the bounding box. Check this out yourself! Also notice that we have used r and R in the definition of the plot region so that if we change the value of either parameter in the first two lines, the plot region will change accordingly.

It is not obvious from the plot that the knot does indeed lie on a torus. It would be great to draw in the torus too. We'll do this in the next section!

Example 6.5.2

```
In[18]:= (* draws the (p,q)-torus knot *)
    r = 1;
    R = 1.5;
    p = 2;
    q = 3;
    knot = ParametricPlot3D[
      {R Cos[p t] + r Cos[p t] Cos[q t],
       R Sin[p t] + r Sin[p t] Cos[q t],
       r Sin[q t]},
      {t, 0, 2 π},
      PlotRange →
        1.1 {{-r - R,  r + R}, {-r - R,  r + R}, {-r, r}},
      PlotStyle → {Tube[.1]}
    ]
```

Out[22]=

6.6 Drawing Graphics Primitives

In the last section, we drew a helix and a torus knot. In both cases it would be nice to also include the cylinder and torus that the curves lie on. We can do this by drawing the cylinder or torus separately and combining the plots.

Example 6.6.1

```
In[34]:= (* plotting graphics primitives *)
    cyl = Graphics3D[
      {
       {Opacity[.2],
        Cylinder[{{0, 0, 0}, {0, 0, 2 π}}, 1]},
       {Red, Thickness[.05],
        Line[{{0, 0, 0}, {0, 0, 2 π}}]}
      }
    ]
```

Out[34]=

The cylinder is a *graphics primitive* much like **Line, Rectangle, Polygon, Circle**, and **Disk** which we already met in Chap. 2. That means we can easily draw a cylinder using the **Graphics3D** function which works in a way entirely analogous to the **Graphics** function which was explained in Chap. 2. Example 6.6.1 draws the cylinder which contains the helix. We have also used the graphics primitive **Line** to draw the axis of the cylinder as well as used a couple of options. We have made the cylinder fairly transparent and also made the axis thicker and **Red**.

The syntax for **Cylinder** is

$$\text{Cylinder}[\{\{x_1, y_1, z_1\}, \{x_2, y_2, z_2\}\}, r]$$

where r is the radius of the cylinder and (x_1, y_1, z_1) and (x_2, y_2, z_2) are the endpoints of its axis.

To combine this with the earlier plot of the helix we now use the **Show** function. This function will display several graphics simultaneously. Notice that when we created the first two plots we named then **helix** and **cyl**. This allows us to refer to them by name in the **Show** function. As usual, we can even add options to **Show**! In this case, we have removed the bounding box and the axes. The results are shown below. We can now clearly see that the helix does indeed lie on the cylinder!

Example 6.6.2

```
In[35]:= (* superimposing helix and cylinder *)
    Show[
      helix, cyl,
      Boxed → False,
      Axes → False
    ]
```

Out[35]=

To insert the torus into our plot of the torus knot we'll first use **RevolutionPlot3D** to plot a torus, name it, and then combine it with the plot of the torus knot using **Show**. We do this in Example 6.6.3. Notice that in describing the generating circle for the surface of revolution, we use the parameters r and R that were already defined when we drew the torus knot. Secondly, just to save space, we have used a semicolon to suppress the output of **RevolutionPlot3D** so we do not see the plot of the torus. (It is worth pointing out that the author first got everything working properly and then added the semicolon at the end.) Finally, we display both plots together using **Show**.

If you rotate the figure by dragging it with the mouse both the knot and the torus move together and it is easy to see how the knot lies on the torus and winds around it 2 times in one direction and 3 times in the other.

Example 6.6.3

```
In[13]:= (* superimposing torus and trefoil knot *)
    torus = RevolutionPlot3D[
        {R + r Cos[t], r Sin[t]}, {t, 0, 2 π},
        PlotRange →
          1.1 {{-r - R, r + R}, {-r - R, r + R},
            {-r, r}},
        Mesh → False,
        PlotStyle → {Opacity[.95]}
        ];
    Show[torus, knot]
```

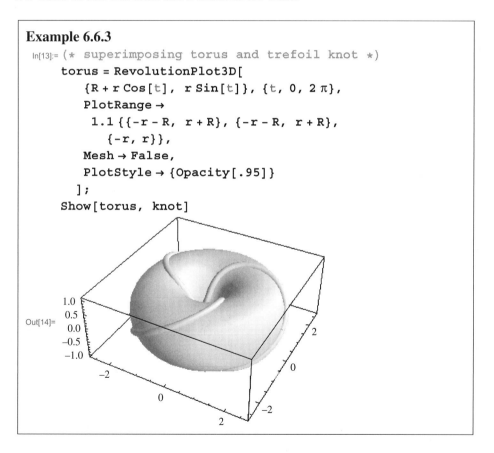

Out[14]=

Using the 3D graphics primitives, it is possible to "build" all sorts of three-dimensional objects. We close this chapter with a single example that hints at the possibilities. In Example 6.6.4, we have drawn a "jungle gym" by using spheres and cylinders. The **Table** function is used 4 times to create the list of all the pieces. The first **Table** creates a list of all the spheres. Each sphere is centered at a point with integer coordinates and has a radius of .25. The **Table** function uses three indices, i, j, and k each of which ranges from -1 to 1.

Example 6.6.4

```
In[33]:= (* a jungle gym built from spheres and
    cylinders *)
  Graphics3D[{
    {Red,
     Table[Sphere[{i, j, k}, .25], {i, -1, 1},
       {j, -1, 1}, {k, -1, 1}]},
    {Blue,
     Table[Cylinder[{{i, j, k}, {i, j, k + 1}},
        .1], {i, -1, 1}, {j, -1, 1}, {k, -1, 0}]},
    {Blue,
     Table[Cylinder[{{i, j, k}, {i, j + 1, k}},
        .1], {i, -1, 1}, {j, -1, 0}, {k, -1, 1}]},
    {Blue,
     Table[Cylinder[{{i, j, k}, {i + 1, j, k}},
        .1], {i, -1, 0}, {j, -1, 1}, {k, -1, 1}]}
  }]
```

Out[33]=

Thus we create a list of 27 spheres. If we want to switch to a different number of spheres, or change the radius of all the spheres, we need only change the arguments in the **Sphere** or **Table** function.

The cylinders that connect the spheres are defined in three sets. Each set is parallel to one of the coordinate axes. Can you see how each **Table** function creates one set of the cylinders? We have run the cylinders all the way into the center of each sphere rather than undertaking the more difficult task of ending them at the surface of the sphere. If you look closely at the graphic you might be able to faintly make out the cylinders extending inside the spheres.

6.7 Find Out More

We have highlighted **Plot3D, RevolutionPlot3D, ParametricPlot3D, Gaphics3D, ContourPlot**, and **ContourPlot3D**, but there are still more *Mathematica* functions that can be used for plotting three-dimensional objects. You might want to look at **SphericalPlot3D** or **RegionPlot3D**. Look these functions up in the Help Files where you can find out what they do, how to use them, and lots of neat examples.

There are also several very nice tutorials that you should look at. As usual, go to the Help Files and search on each of the following to find the tutorial:

- tutorial/ThreeDimensionalSurfacePlots
- tutorial/ThreeDimensionalGraphicsPrimitives
- tutorial/DensityAndContourPlots
- tutorial/LightingAndSurfaceProperties

Quiz

1. Use **Plot3D** to plot the graph of the function $\cos x + \sin y$. Let both x and y run from zero to 4π.

2. Repeat the first plot, but only graph that part of the surface that lies above the xy-plane. (Hint: Use **RegionFunction**.)

3. Modify the previous plot still further so that only one "bump" of the surface is displayed as shown below. Use **BoxRatios** so that it is drawn with the correct aspect ratio.

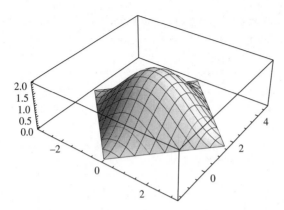

4. Modify the plot even further so as to cut out a quarter of the bump as shown below.

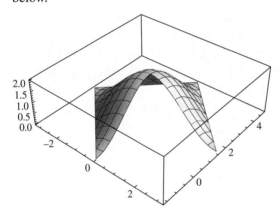

5. Plot the contour lines of $\cos x + \sin y$.

6. Create a "torus with square cross section" by using a square as the generating curve of a surface of revolution.

7. Repeat the previous exercise, but use **Manipulate** to allow the square to be rotated around its center. (This is challenging!)

8. Draw a right circular cone whose base is a disk in the xy-plane with radius 1 and center at $(0, 0)$ and whose cone point is located at $(0, 0, 5)$. (Hint: Use **RevolutionPlot3D** or **RegionPlot3D**.)

9. Repeat the previous exercise, but use **Manipulate** to allow the cone point to be moved around in space. (Warning: This is hard!)

10. Use **ParametricPlot3D** to draw *Lissajous knots*. These are knots of the form

$$x(t) = \cos(n_x t + \phi_x)$$
$$y(t) = \cos(n_y t + \phi_y)$$
$$z(t) = \cos(n_z t + \phi_z)$$

Start with $n_x = 3$, $n_y = 5$, $n_z = 7$, $\phi_x = 0$, $\phi_y = \pi/4$, and $\phi_z = \pi/12$. Use the **Tube** option to make your plot look really nice!

CHAPTER 7

Calculus

If mathematics is the language of science, then calculus is a large part of the dictionary. Calculus is exactly what is needed to describe how things change, and describing how things change is an important way in which scientists use mathematics. The authors of the book *Calculus in Context*[1] looked at how scientists actually use calculus in their work and discovered (not surprisingly) that modeling with differential equations is one of the primary uses.

Mathematica can be used for all the usual computations involving calculus: taking limits, finding derivatives, and computing integrals. In this chapter we'll try to hit the highlights of these computations. Obviously, this chapter cannot substitute for a course in calculus. Our goal is to not to teach you calculus (we assume you already know some, or are currently taking a course), but rather to describe a handful of useful *Mathematica* functions.

7.1 Limits

The really big idea of calculus is that of the *limit*. Once this notion is defined and understood it leads to the development of both the derivative as well as the integral, the two pillars of calculus.

---

[1] In the humble opinion of the author, this book by J. Callahan, D.A. Cox, K.R. Hoffman, D. O'Shea, H. Pollatsek, and L. Senechal (Freeman, 1995) is perhaps the most innovative calculus textbook of the late twentieth century.

Suppose that $f(x)$ is some function of x and we would like to know the behavior of $f(x)$ as x approaches some specific point, say x_0. Of course, the value $f(x_0)$ of $f(x)$ at x_0 is important for our understanding of f, but the value of $f(x)$ at x_0 may or may not exist, and even if it does, it might be wholly unrelated to the behavior of f near x_0. The *limit of $f(x)$ as x approaches x_0* is designed to capture information about this behavior. However, the limit also may or may not exist, and even if it does, might be unrelated to the value of $f(x)$ at x_0. (Of course, for nice functions, both the limit of the function and the value of the function will exist at x_0 and furthermore, be equal to each other. In this case, the function is called *continuous at x_0*. But we are considering the most general possible case here.)

Recall that the limit of $f(x)$ as x approaches x_0, should it exist, is the number L such that the difference between $f(x)$ and L can be made arbitrarily small by choosing x sufficiently close to, but not equal to, x_0. If the limit does exist, and is equal to L, we'll write

$$L = \lim_{x \to x_0} f(x)$$

Let's look at some examples.

A classic example involves $f(x) = \sin x / x$. Notice that $f(x)$ does not exist at zero. If we try to substitute zero for x this gives the meaningless expression $0/0$. But, remember that having a value at a point and having a limit at a point can be entirely different things. Perhaps this function still has a limit as x approaches 0. Consider the following *Mathematica* example.

Example 7.1.1

```
In[35]:= (* computing Sin[x]/x near zero *)
         f[x_] := Sin[x]
                  ------
                    x
         f[0]
         N@Table[f[10^k], {k, 0, -5, -1}]
         N@Table[f[-10^k], {k, 0, -5, -1}]
```

Power::infy : Infinite expression $\frac{1}{0}$ encountered. »

∞::indet : Indeterminate expression 0 ComplexInfinity encountered. »

Out[36]= Indeterminate

Out[37]= {0.841471, 0.998334, 0.999983, 1., 1., 1.}

Out[38]= {0.841471, 0.998334, 0.999983, 1., 1., 1.}

We first define $f(x) = \sin x / x$ and then try to evaluate $f(0)$, which fails. Next we use **Table** to evaluate $f(x)$ at increasingly smaller positive numbers. Note the use of the numerical function **N** and the construction **N@Table[]**. In general, if *f* is a function, then *f* @ *expr* is equivalent to *f*[*expr*]. The values of the function are getting closer and closer to 1 — eventually so close that *Mathematica* displays the values as 1. We repeat the experiment for negative numbers that are approaching zero and, of course, get the same results since $f(-x) = f(x)$. Based on this experiment it certainly seems like the limit of $f(x)$ as x approaches zero is 1, and, in fact, it is well known that this is the case.[2]

As further evidence of this, let's plot the graph of $f(x)$.

Example 7.1.2

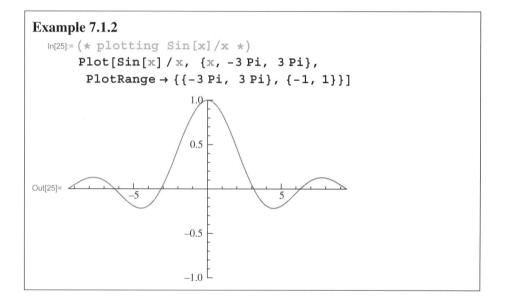

```
In[25]:= (* plotting Sin[x]/x *)
        Plot[Sin[x] / x, {x, -3 Pi, 3 Pi},
          PlotRange → {{-3 Pi, 3 Pi}, {-1, 1}}]
```

---

[2]It is not too hard to prove this. Let x be a small angle and consider the circular sector with radius 1 and central angle x. Triangle OBC has more area then the sector which in turn has more area then the triangle OAD. This gives the inequality

$$\frac{\tan x}{2} > \frac{x}{2} > \frac{\sin x \cos x}{2}$$

which can be rewritten as

$$\frac{1}{\cos x} > \frac{x}{\sin x} > \cos x.$$

Thus as x approaches 1, $\frac{x}{\sin x}$ is trapped between two quantities that are each approaching 1, so must approach 1 itself.

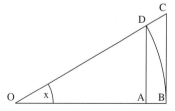

Notice that *Mathematica* makes no objection to drawing the graph even though the function does not exist at zero. Technically, the graph of $f(x)$ should contain a "hole" at the point $(0, 1)$, but, of course, this is impossible to discern from the plot.

Finally, it turns out that we can just ask *Mathematica* directly for the limit! The function **Limit** is just what we need and is illustrated in the next example.

Example 7.1.3

In[9]:= `(* finding a limit *)`
`Limit[Sin[x] / x, x → 0]`

Out[9]= 1

Note that **Limit** takes two arguments: the expression whose limit we wish to evaluate as x approaches x_0 and then the argument $\mathbf{x} \rightarrow \mathbf{x0}$ to indicate that x should approach x_0. Here are some more examples.

Example 7.1.4

In[16]:= `(* a handful of limits *)`

$$\text{Limit}\left[\frac{e^x}{x}, \; x \rightarrow \text{Infinity}\right]$$

$$\text{Limit}\left[\frac{\text{Log}[x]}{x}, \; x \rightarrow \text{Infinity}\right]$$

$$\text{Limit}\left[\frac{ax^2 + b}{cx^2 + dx + e}, \; x \rightarrow \text{Infinity}\right]$$

$$\text{Limit}\left[\sqrt{x} - \sqrt{x-1}, \; x \rightarrow \text{Infinity}\right]$$

$$\text{Limit}\left[\left(1 + \frac{a}{x}\right)^x, \; x \rightarrow \text{Infinity}\right]$$

$$\text{Limit}\left[\left(\frac{\text{Sin}[x]}{x}\right)^{\frac{1}{x^2}}, \; x \rightarrow 0\right]$$

Out[16]= ∞

Out[17]= 0

Out[18]= $\dfrac{a}{c}$

Out[19]= 0

Out[20]= e^a

Out[21]= $\dfrac{1}{e^{1/6}}$

In Example 7.1.3, we are led to the *indeterminant* expression 0/0 if we try to substitute zero for x. Other possible indeterminant expressions are

$$0 \cdot \infty, \ \infty - \infty, \ \frac{\infty}{\infty}, \ 1^{\infty}, \ 0^{0}, \text{ and } \infty^{0}$$

Each of the limits in Example 7.1.4 is associated to one of these indeterminant forms. Notice that we may take limits as x approaches infinity, and that even if a limit does not exist it still might make sense to declare the limit to be infinity (or negative infinity). **Limit** is also capable of dealing with expressions that have unknown constants and parameters and simply returns an answer that involves the parameters. This is the case in the third and fifth limits in the above example.

7.2 One-Sided Limits

Consider the graph of $f(x) = \tan x$, whose graph is shown below between -2π and 2π.

We have plotted the function twice to point out a subtlety that can occur when a function is not continuous. Notice that in the first plot vertical lines appear at odd multiples of $\pi/2$. These are not part of the graph and should not be present. To eliminate them we use the **Exclusions** option in the second plot and specifically tell *Mathematica* not to plot anything at points where $\tan x$ is undefined. Notice that we use the **Table** function to create the list of points to be excluded from the plot.

What is the limit of $\tan x$ as x approaches $\pi/2$? In this case, it depends on whether we approach $\pi/2$ from the right or from the left. If we approach from the right the function approaches negative infinity, but if we approach from the left, the function approaches positive infinity. Thus different *one-sided limits* exist at $\pi/2$. We can

---

Example 7.2.1

```
In[35]:= (* using exclusion in a plot *)
    Plot[Tan[x], {x, -2 π, 2 π},
     Ticks → {Table[k π / 2, {k, -4, 4}], None}
    ]
    Plot[Tan[x], {x, -2 π, 2 π},
     Ticks → {Table[k π / 2, {k, -4, 4}], None},
     Exclusions → Table[j π / 2, {j, -3, 3, 2}]
    ]
```

Example 7.2.1 (Continued)

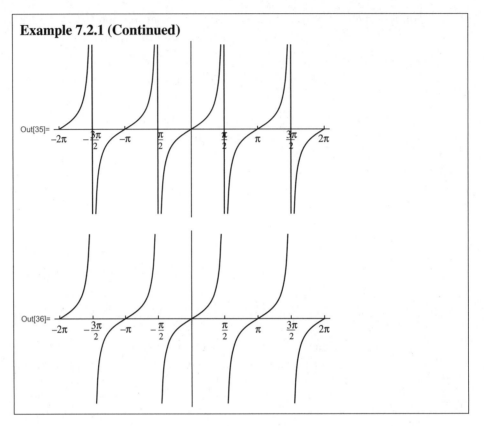

restrict the **Limit** function to take only one-sided limits as shown in Example 7.2.2. To do this, we use the **Direction** option. Using this option with a value of 1 or −1 will produce one-sided limits from either the left or right side, respectively.

Example 7.2.2

```
In[37]:= (* finding one-sided limits *)
        Limit[Tan[x], x → π / 2, Direction → 1]
        Limit[Tan[x], x → π / 2, Direction → -1]

Out[37]= ∞

Out[38]= − ∞
```

Sometimes a function just doesn't have a limit at a point, even one-sided limits. A classic example involves the function $f(x) = \sin(1/x)$, which has no limit at zero. The behavior of $f(x)$ near zero is quite interesting, and we plot the graph of $f(x)$ below.

Example 7.2.3

In[41]:= (* a function with no limit at zero *)
     Plot[Sin[1 / x], {x, -1, 1}]

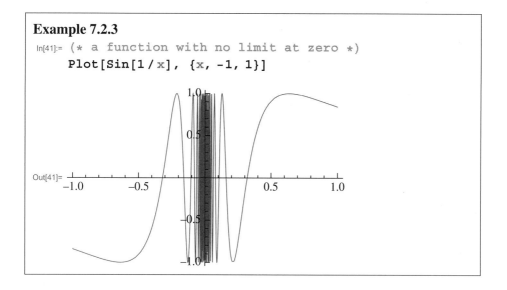

As x approaches zero the function oscillates up and down between 1 and -1 with ever increasing frequency and does not approach any limit. What does **Limit** give in this case?

Example 7.2.4

In[43]:= (* there is no limit at zero *)
     Limit[Sin[1 / x], x → 0]

Out[43]= Interval[{-1, 1}]

As we see in the example, *Mathematica* can only report that the function stays trapped in the interval from -1 to 1 as x approaches zero.

7.3 Multivariable Limits

Sometimes we want to find the limit of a function of several variables. For example, what is

$$\lim_{(x,y)\to(0,0)} \frac{x^2 - y^2}{x^2 y^4 + x^4 y^2}$$

Of course, right away we see that the function is not defined at $(0, 0)$. In fact, it is not defined on either coordinate axis. To try to see what is going on, we have several tools at our disposal: we can use **Plot3D** to visualize the graph of the function, we can use **ContourPlot** to also help see what the graph must look like, and we can also use **Limit** if we restrict ourselves to approaching $(0, 0)$ along various curves. Unfortunately, *Mathematica* does not have any function specifically designed for computing multivariable limits. Let's start with **Plot3D** and **ContourPlot**.

Example 7.3.1

In[44]:= **Plot3D** $\left[\dfrac{x^2 - y^2}{x^2 y^4 + x^4 y^2} , \{x, -1, 1\}, \{y, -1, 1\} \right]$

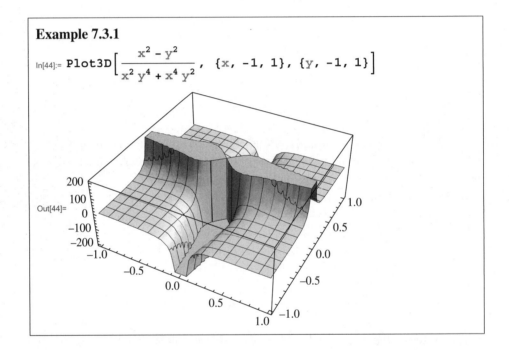

Out[44]=

From the plot we can see that something nasty appears to be happening near zero! Away from $(0, 0)$, as we approach the x-axis the function appears to be approaching infinity and as we approach the y-axis the function appears to be approaching negative infinity. Example 7.3.2 gives us the contour plot.

From the contour plot it appears that the lines $y = \pm x$ are contours, or level curves. Sure enough, if we let $y = \pm x$ then the function becomes $0/(2x^6) = 0$ for all $x \neq 0$. So, if we were to approach $(0, 0)$ along the line $y = x$ or along the line $y = -x$, the limit would be zero. But what if we approach $(0, 0)$ along other lines, say the line $y = mx$? We can use **Limit** to do this as the next example shows.

Example 7.3.2

In[45]:= `ContourPlot`$\left[\dfrac{x^2 - y^2}{x^2\,y^4 + x^4\,y^2},\ \{x,\ -1,\ 1\},\ \{y,\ -1,\ 1\}\right]$

Out[45]=

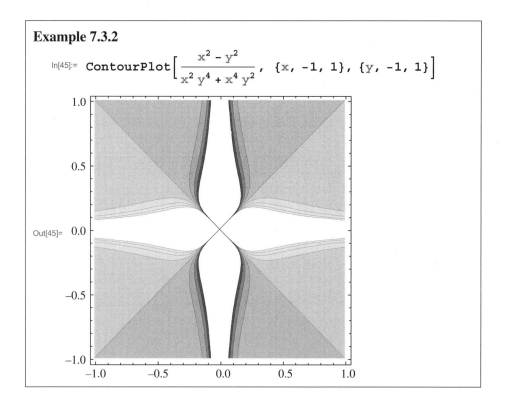

Example 7.3.3

In[47]:= `(* approaching the origin along the line y=mx *)`

`Limit`$\left[\dfrac{x^2 - y^2}{x^2\,y^4 + x^4\,y^2}\ \text{/. } y \to m\,x,\ x \to 0\right]$

Out[47]= `DirectedInfinity`$\left[\dfrac{\text{Sign}\left[1 - m^2\right]}{\text{Sign}[m]^2\,\text{Sign}\left[1 + m^2\right]}\right]$

Here we have used the replacement rule /.**y** → **m x** to first substitute $y = mx$ into the expression before taking the limit as x approaches zero. We get a somewhat strange answer involving **DirectedInfinity** so we need to know what this function is in order to understand what we got. If z is a nonzero point in the complex plane, then **DirectedInfinity[z]** is best thought of as the point at infinity obtained by starting at zero and heading off to infinity in the direction of z. Thus **DirectedInfinity[1]** is ∞, while **DirectedInfinity[−1]** is $−\infty$. In this case, we see that the argument of the **DirectedInfinity** function is itself a function of the slope m, which makes

Mathematica Demystified

sense. The **Sign[x]** function is defined to be 1 if $x > 0$, -1 if $x < 0$ and 0 if $x = 0$. Let's plot the argument of the **DirectedInfinity** function.

Example 7.3.4

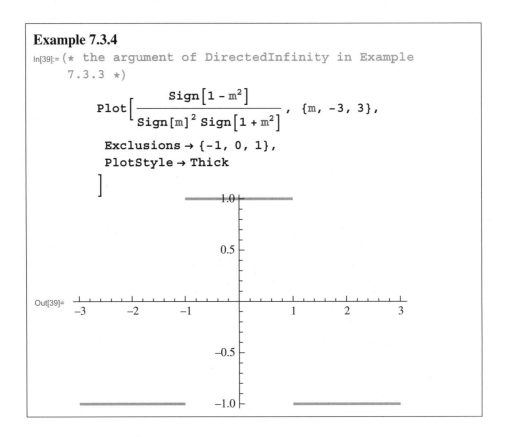

We see that if m is between -1 and 1 (and excluding zero), then the limit of our original function is **DirectedInfinity[1]** which is infinity. But if $|m| > 1$ then the limit is **DirectedInfinity[−1]** which is negative infinity. And, as we first saw, if $m = \pm 1$ the limit is zero. The upshot of our investigations is that the $f(x)$ does not have a limit at x approaches $(0, 0)$. If it did have a limit, L, we would always get L no matter how we approach $(0, 0)$. But what we have discovered is that approaching $(0, 0)$ along different routes (in this case, straight lines) yields different limits.

Recall that a function is *continuous* at a point if it has a value there and a limit there and the two are equal. Graphically, places where a function is not continuous show up as breaks or tears in the graph, or places where the graph does not exist. The function $\sin(x)/x$ does not exist at zero so it is not continuous there and its graph has a hole in it. The graph of $\tan x$ is even worse at the places where $\tan x$ does not exist: instead of having holes, it has huge "jumps" where it instantly jumps from infinity to negative infinity. Our last example, $\frac{x^2 - y^2}{x^2 y^4 + x^4 y^2}$ does not exist at $(0, 0)$ and so is not continuous there. Its graph is very badly behaved near $(0, 0)$!

7.4 Derivatives

Given a function $f(x)$ its *derivative* can be defined as the limit

$$\lim_{h \to 0} \frac{f(x + h) - f(x)}{h}$$

The derivative can be interpreted as the instantaneous rate of change of $f(x)$. Geometrically, the derivative gives the slope of the line tangent to the graph of $f(x)$. Of course, as we have seen in the first part of this chapter, limits do not always exist, so sometimes a function may not be *differentiable*, that is to say, have a derivative. Of course, all "nice" functions are differentiable. In particular, all polynomials are, the exponential and logarithmic functions are and so are all the trigonometric functions. If we then start combining differentiable functions by adding, subtracting, multiplying, dividing, or composing we will produce differentiable functions, with some exceptions occurring if we try to do unacceptable things like divide by zero.

We can take the derivative of a function in *Mathematica* by using the derivative function **D**. For example, suppose that $f(x)$ is the polynomial $f(x) = x^4 - x^3 + 2x + 1$. The **D** function takes two arguments. The first is the function we want to differentiate and the second is the variable with respect to which we want to differentiate. Example 7.4.1 illustrates this.

Example 7.4.1

In[48]:= (* differentiating a polynomial *)
    D[x^4 - x^3 + 2 x + 1, x]

Out[48]= $2 - 3 x^2 + 4 x^3$

Of course, differentiating polynomials is pretty easy to do in our head. But imagine trying to differentiate the function in the next example in your head!

Mathematica Demystified

Example 7.4.2

In[58]:= (* a complicated derivative *)

$$D\left[ArcTan\left[Sin\left[x^{10}\right]\right] e^{1/x}, x\right]$$

Out[58]= $-\dfrac{e^{\frac{1}{x}} ArcTan\left[Sin\left[x^{10}\right]\right]}{x^2} + \dfrac{10\, e^{\frac{1}{x}} x^9\, Cos\left[x^{10}\right]}{1 + Sin\left[x^{10}\right]^2}$

An alternative to using **D** is to use the "prime" notation for derivatives introduced by Lagrange.[3] Example 7.4.3 illustrates this and also the fact that there appears to be no limit as to how many primes we can use to denote higher and higher derivatives. We also plot the function and its first and second derivative.

Example 7.4.3

In[67]:= (* using the "prime" notation for D *)

```
f[x_] := Exp[x] + Sin[x]
f'[x]
f''[x]
f'''[x]
f''''[x]
f'''''[x]
Plot[{f[x], f'[x], f''[x]}, {x, -3, 1},
  PlotStyle → Thick]
```

Out[68]= $e^x + Cos[x]$

Out[69]= $e^x - Sin[x]$

Out[70]= $e^x - Cos[x]$

Out[71]= $e^x + Sin[x]$

Out[72]= $e^x + Cos[x]$

---

[3] Joseph-Louis Lagrange (1736–1813) was an important mathematician. It is he who introduced the notation $f'(x)$ (also $f''(x)$ etc.) for the derivative of $f(x)$ with respect to x. The purpose of the notation, he says, is to free the intellect from the false idea of the infinitely small (as Leibniz's notation $\frac{dy}{dx}$ does not free the intellect), and to make clear that the derivative is a function just like $f(x)$, and derived from $f(x)$. The term "fonction dérivée" from which the English term "derivative" comes is also due to Lagrange. He first introduced the prime notation in the 1770s but it was done systematically in his *Théorie des fonctions analytiques* of 1797.

Example 7.4.3 (Continued)

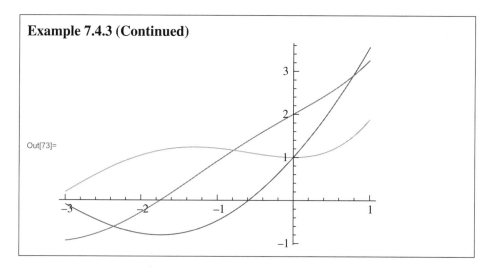

Out[73]=

The prime notation is handy, but clearly not what we want to use for higher order derivatives. For anything past the third derivative it is probably better to switch back to the **D** function with the optional syntax **D[f[x], {x, n}]** for taking the *n*th derivative. For example, suppose we want to find the first, through fifth derivatives of tan *x*. Example 7.4.4 does this. Note that we have used **Table** to make a table of the successive derivatives. Each entry in the table is of the form **D[Tan[x], {x, n}]**. We have also used **TableForm** with the @ construction to display the elements of the table nicely.

Example 7.4.4

```
In[74]:= (* the first 5 derivatives of Tan[x] *)
       TableForm@Table[D[Tan[x], {x, n}], {n, 0, 5}]
Out[74]//TableForm=
       Tan[x]
       Sec[x]²
       2 Sec[x]² Tan[x]
       2 Sec[x]⁴ + 4 Sec[x]² Tan[x]²
       16 Sec[x]⁴ Tan[x] + 8 Sec[x]² Tan[x]³
       16 Sec[x]⁶ + 88 Sec[x]⁴ Tan[x]² + 16 Sec[x]² Tan[x]⁴
```

If we want to take partial derivatives of a function of more than one variable, we simply list the variable that we wish to differentiate with respect to. To take multiple partial derivatives, simply list the variables in the order of the differentiation. Example 7.4.5 should make this clear.

Example 7.4.5

In[75]:= (* examples of partial derivatives *)
      D[Exp[3 x^2 + y], x]
      D[Sin[x y], x, y]
      D[g[x y], {y, 2}]

Out[75]= $6 e^{3 x^2 + y} x$

Out[76]= $Cos[x y] - x y Sin[x y]$

Out[77]= $x^2 g''[x y]$

Here we have taken the partial derivative of e^{3x^2+y} with respect to x, the second partial derivative of $\sin(xy)$, first with respect to x and then with respect to y, and finally, the second partial derivative of g with respect to y. When taking partial derivatives *Mathematica* treats all variables other than the variable of differentiation as constants. Notice that in the last example the function g was not previously defined so *Mathematica* treated it abstractly.

We close this section by pointing out an easy to make error that illustrates some subtle issues in defining functions as well as a subtle difference between **D[f[x], x]** and **f'[x]**. Suppose we define a function and then want to name its derivative so that we can then use the derivative for things, including plotting the graph of the derivative. Example 7.4.6 does not plot the derivative! But if the second line is changed to **g[x_] := f'[x]** it will plot the derivative! (Try it!!) What is going on here?

Example 7.4.6

In[39]:= (* this will NOT plot the derivative! *)
       (* a problem due to delayed definition *)
      f[x_] := x^3
      g[x_] := D[f[x], x]
      Plot[g[x], {x, -1, 1}]

General::ivar : −0.999959 is not a valid variable. ≫

General::ivar : −0.959143 is not a valid variable. ≫

General::ivar : −0.918326 is not a valid variable. ≫

General::stop : Further output of General ::ivar will be suppressed during this calculation. ≫

Example 7.4.6 (Continued)

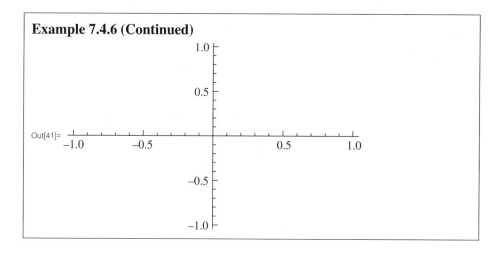

The problem above occurs when we try to plot the derivative. **Plot** attempts to evaluate the expression **g[x]** at various points between -1 and 1 so that it can then plot these values. We defined g using the delayed definition construction (using $:=$) rather than an immediate definition construction (using $=$). Recall that we discussed the two methods of defining a function in Chap. 5. Since we defined g using delayed definition, the definition of g as the derivative of f is substituted at the time that g is called. The upshot of all of this is that when **Plot** tries to evaluate g at $x = -.999959$ it substitutes that value into both occurrences of x in **D[f[x], x]**, after which *Mathematica* is trying to take the derivative of $f(-.999959)$ with respect to the "variable" $-.999959$ which, of course, is not a variable. Hence the error message. A similar problem will happen with the instruction **Table[D[f[x], x], {x, 1, 10}]**. In this case, *Mathematica* will complain that "1 is not a variable," "2 is not a variable," and so on.

There are several ways to solve the problem. The first way is to use immediate, rather than delayed definition when we define g. If we change the second line to **g[x] = D[f[x], x]** then the moment that g is defined it will be defined as $3x^2$ (because this is the derivative of x^3 with respect to x). Then when g is evaluated by **Plot** it will be $3x^2$ that is evaluated, not **D[f[x], x]**. This is one case where it make sense to use immediate rather than delayed definition to define a function.

Another solution is to keep g defined with delayed definition but use a "dummy" variable when we try to evaluate g in the **Plot** function. If we change the third line to **Plot[g[t]/.t $\rightarrow$ x, {x, -1, 1}]** it will work. In this case when **Plot** encounters **g[t]** it will substitute **D[f[t], t]**. This will be computed as $3t^2$ and then evaluated at various values of x between -1 and 1 by replacing t with x.

This is really what is happening if we go with the third solution of defining g as $\mathbf{g[x\_]} := \mathbf{f'[x]}$. This is because *Mathematica* defines $\mathbf{f'[x]}$ as $\mathbf{D[f[t], t]/.t \to x}$. (We could also define g immediately as $\mathbf{g[x\_] = f'[x]}$ and this will work too.)

7.5 Minimum/Maximum Problems

One of the practical applications of calculus is to find maxima and minima. For example, suppose we want to find the shape of a cylindrical can that contains the most volume for a given amount of surface area. Suppose the can has a circular base of radius r and a height of h. Then its volume is $V = \pi r^2 h$ (area of the base times the height) while its surface area is $S = 2\pi r^2 + 2\pi rh$ (twice the area of the base plus the area of the sides, which is the circumference times the height). Keeping S fixed we want to vary r and h so as to obtain the largest volume. The volume is ostensibly a function of two variables, but because the surface area must remain constant the two variables are not independent. In fact, we can solve for one in terms of the other and the surface area S. It is easiest to solve for h in terms of r and S obtaining $h = \frac{S}{2\pi r} - r$. If we now substitute this into the formula for the volume we obtain $V = \pi r^2 \left(\frac{S}{2\pi r} - r\right) = \frac{Sr}{2} - \pi r^3$. This is a cubic function of r and we have plotted its graph in Example 7.5.1. Note that we have let $S = 1$ in the plot, which makes sense because we may as well assume there is "one unit" of area.

We now have the volume as a function of r alone (and the constant S) and we want to know what value of r makes this as large as possible. Of course, r cannot be any real number. It makes no sense for r to be negative and it also cannot be the case that r is too big. For as r gets bigger the area of the top and bottom of the can alone will soon add up to more than S. In fact, the biggest that r can be is $\sqrt{\frac{S}{2\pi}}$ at which point the top and bottom will each have area $S/2$ and there will be no area left for the sides! At this point the height is zero and so is the volume. In our graph, the cubic must cross the r-axis at $\sqrt{\frac{S}{2\pi}} = \sqrt{\frac{1}{2\pi}} \simeq 0.398942$ since we have let $S = 1$ in the plot.

At the other extreme is $r = 0$, where again we have a volume of zero. In order to get a can with nonzero volume, we should assume that r lies strictly between zero and $\sqrt{\frac{S}{2\pi}}$. Thus the only portion of the curve that we just plotted that we want to consider is the portion in the first quadrant where both the radius and the volume are positive. Looking at the graph, it is clear that it has a unique maximum value on this domain.

We can find the maximum by using the fact that at the maximum the derivative is zero, or in graphical terms, the slope is zero, which is to say, the tangent line

---

Example 7.5.1

In[40]:= (* the volume of a can with one unit of
        surface area as a function of its radius *)

$$\text{Plot}\left[\frac{r}{2} - \pi\, r^3,\ \{r,\ -.5,\ .5\}\right]$$

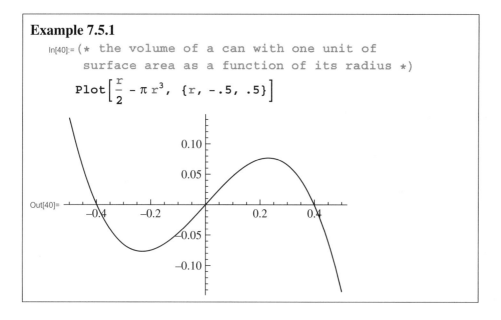

Out[40]=

---

is horizontal. So, to find the maximum we simply need to compute the derivative and then find out where the derivative is zero. All of these steps are contained in Example 7.5.2.

In the first cell, we define the volume and area. Next we use the **Solve** function to solve for the height in terms of the radius and the area S and then substitute this expression of h into the volume formula.[4]

We then take the derivative of the volume with respect to the radius and simplify what we get. Finally, we set the derivative equal to zero and solve for the radius, obtaining two solutions. Only one is positive, so we let r equal that value. Going back to the area equation, we use **Solve** again to find the height.

In the last cell, we take the ratio of the height to twice the radius (i.e., the ratio of the height to the diameter) and get a value of 1. Thus the can with the largest volume for a given surface area is the can that has the same height as diameter. That is, a vertical cross containing the axis of the can is a square. Taking a quick look at all the cans that are in my kitchen pantry I don't see one can with this shape! Why aren't can manufacturers making efficient cans? What other factors in can manufacturing, packing, shipping, and so on, have we left out that might lead to a different optimal solution?

---

[4]We describe how to use the **Solve** function in detail in Chap. 8.

Example 7.5.2

In[20]:= `(* finding max volume of can with surface area S *)`
`volume = Pi r^2 h`
`area = 2 Pi r^2 + 2 Pi r h`

Out[20]= $h \pi r^2$

Out[21]= $2 h \pi r + 2 \pi r^2$

In[22]:= `(* find height in terms of radius *)`
`Solve[area == S, h]`

Out[22]= $\left\{\left\{h \to \dfrac{-2 \pi r^2 + S}{2 \pi r}\right\}\right\}$

In[23]:= `volume /. %` `(* volume in terms of r and S *)`

Out[23]= $\left\{\dfrac{1}{2} r \left(-2 \pi r^2 + S\right)\right\}$

In[24]:= `(* derivative of volume wrt to r *)`
`D[%, r]`

Out[24]= $\left\{-2 \pi r^2 + \dfrac{1}{2} \left(-2 \pi r^2 + S\right)\right\}$

In[25]:= `Simplify[%]`

Out[25]= $\left\{\dfrac{1}{2} \left(-6 \pi r^2 + S\right)\right\}$

In[26]:= `(* find r where slope is zero *)`
`Solve[% == 0, r]`

Out[26]= $\left\{\left\{r \to -\dfrac{\sqrt{S}}{\sqrt{6 \pi}}\right\}, \left\{r \to \dfrac{\sqrt{S}}{\sqrt{6 \pi}}\right\}\right\}$

In[27]:= `r = r /. %[[2]]` `(* use positive answer *)`

Out[27]= $\dfrac{\sqrt{S}}{\sqrt{6 \pi}}$

In[29]:= `h = h /. Solve[area == S, h]` `(* find height *)`

Out[29]= $\left\{\sqrt{\dfrac{2}{3 \pi}} \sqrt{S}\right\}$

In[30]:= `(* find ratio of height to diameter *)`
`h / 2 / r`

Out[30]= $\{1\}$

As is often the case, *Mathematica* has a built in function that is perfect for the problem we just did! The function **FindMaximum** will find the maximum of a function provided that we provide an initial guess that is near where the maximum occurs. Let's use **FindMaximum** to find the maximum of the curve we plotted in Example 7.5.1. By looking at the graph, we can easily estimate the maximum to be near $x = 0.2$ so we will use that as our initial guess. Example 7.5.3 illustrates this, as well as what happens if we start with other initial guesses. Note that we begin with **Clear[r]** because in Example 7.5.2 we gave r the value $\sqrt{S}/\sqrt{6\pi}$ and we now need r to be a variable.

Example 7.5.3

In[41]:= (* using FindMaximum with different initial
    guesses *)
Clear[r]
FindMaximum $\left[\frac{r}{2} - \pi r^3, \{r, 0.2\}\right]$

FindMaximum $\left[\frac{r}{2} - \pi r^3, \{r, 0\}\right]$

FindMaximum $\left[\frac{r}{2} - \pi r^3, \{r, 1\}\right]$

FindMaximum $\left[\frac{r}{2} - \pi r^3, \{r, -1\}\right]$

Out[42]= {0.0767765, {r → 0.230329}}

Out[43]= {0.0767765, {r → 0.230329}}

Out[44]= {0.0767765, {r → 0.230329}}

    FindMaximum::cvmit : Failed to converge to the requested accuracy or precision within 100 iterations. »

Out[45]= $\left\{6.577266015991333 \times 10^{313}, \{r → -2.75612 \times 10^{104}\}\right\}$

To use **FindMaximum**, we provide two arguments. The first is the function that we want to maximize. The second is a list of two elements, the first being the variable and the second being our initial guess. As you can see in the example, if the initial guess is not close enough to the maximum (as -1 was not in this case), then the function can fail. When it does work, it returns the maximum value of the function followed by the point where it occurs.

Mathematica also provides the function **FindMinimum** which is similar to **FindMaximum** but which finds minima instead. Both of these functions will also work with functions of several variables. They both attempt to find *local* extrema (either minima or maxima) near the initial guess provided by the user. Closely related to these functions are the functions **Maximize** and **Minimize** which attempt to find *global* extrema. Moreover, these latter two functions can work symbolically. All four of these functions can also handle *constraints*. Let's see how to redo the can problem by using constraints.

The volume of the can is given by $V = \pi r^2 h$ which, as we already mentioned, is a function of two variables. Our initial technique for maximizing this was to recognize that the variables are not independent since the surface area is fixed. This allowed us to turn the problem into a single-variable problem. But we don't need to do this! What we really want to do is maximize the volume subject to the constraint that the surface area is fixed. *Mathematica* will let us do this directly as is seen in the following example. Here we have used **FindMaximum** as before, but this time entered as our first argument a list that contains as its first element the function we want to maximize and as its second element the constraint. For our initial guess we now provide a list of each variable with its initial guess.

Example 7.5.4

In[46]:= (* solving the can problem using FindMaximum
        with constraint *)
    Clear[r, h]
    volume = $\pi\, r^2\, h$;
    area = $2\,\pi\, r\, h + 2\,\pi\, r^2$;
    FindMaximum[{volume, area == 1}, {{r, 1}, {h, 1}}]

Out[49]= {0.0767765, {r → 0.230329, h → 0.460659}}

Note that **FindMaximum** and **FindMinimum** will not work with symbolic expressions. So we cannot enter the constraint in the last example as **area == S**, where **S** is an unknown constant. However, **Maximize** and **Minimize** can handle symbolic expressions. Try it out! (Be warned however that the answer *Mathematica* gives might be kind of complicated. You may need to dig into the Help Files to see how to interpret the answer!)

As a final example, suppose we want to find the point on the ellipsoid $x^2 + 2y^2 + 6z^2 = 1$ that is closest to the point $(4, 6, 13)$. If (x, y, z) is any point in space, we'll minimize its distance to $(4, 6, 13)$ subject to the constraint that it lie on the ellipsoid. Actually, we'll minimize the square of the distance, which will still find the closest point, but allow for a slightly simpler function to be minimized. The next example gives the result, as well as the point on the ellipsoid that is farthest away!

Here the square of the distance from the point (x, y, z) to the point $(4, 6, 13)$ is given by $(x - 4)^2 + (y - 6)^2 + (z - 13)^2$. This is the function we want to minimize (or maximize). The constraint is the fact that (x, y, z) must lie on the ellipsoid, or in other words, $x^2 + 2y^2 + 6z^2 = 1$.

Example 7.5.5

In[50]:= (* finding the closest and farthest points
on an ellipsoid to (4,6,13) *)
Minimize$\Big[$
    $\{(x-4.0)^2 + (y-6)^2 + (z-13)^2,\ x^2 + 2\,y^2 + 6\,z^2 == 1\}$,
    $\{x,\ y,\ z\}$
$\Big]$

Maximize$\Big[$
    $\{(x-4.0)^2 + (y-6)^2 + (z-13)^2,\ x^2 + 2\,y^2 + 6\,z^2 == 1\}$,
    $\{x,\ y,\ z\}$
$\Big]$

Out[50]= $\{205.694,\ \{x \to 0.474615,\ y \to 0.378411,\ z \to 0.285292\}\}$

Out[51]= $\{237.262,$
$\{x \to -0.541677,\ y \to -0.380483,\ z \to -0.263645\}\}$

7.6 Series

Approximating a function by polynomials is quite useful and leads naturally to the theory of infinite series. Suppose we are considering a function $f(x)$ and would like to find a polynomial that closely approximates $f(x)$ over some domain. Of course, if f itself is already a polynomial, then there is not much to do! So we should be thinking that f is not a polynomial. Perhaps f is a trigonometric function, or a logarithm or some other interesting function. Finding a polynomial that closely approximates f will be useful because polynomials are easy to compute and work with.

What does it mean for a polynomial $P(x)$ to approximate $f(x)$? It is probably unrealistic to hope that P will be close to f for all values of x, so we mean that near some point the two functions are nearly equal. So, let's pick a point, say $x = a$, and try to approximate f near a. Since we want the functions to be nearly the same near a, it makes sense to demand that P and f are exactly the same at a, that is $P(a) = f(a)$. This means that the two graphs will intersect at $x = a$. But we want the graphs to be nearly the same, so let's make them tangent to each other at a as well. This means that $P'(a) = f'(a)$. Similarly, we would like the graphs to be bending the same way at a, so let's also demand that the second derivatives of P and f also agree at a. A good idea is to simply decide that P should share the same value, first derivative, second derivative, third derivative, and so on, with f at a.

Of course, there is a limit to how long this can go on, because the higher order derivatives of any polynomial are eventually zero.

Suppose

$$P(x) = a_0 + a_1(x - a) + a_2(x - a)^2 + \cdots + a_n(x - a)^n$$

We have expressed the polynomial as a sum of powers of $x - a$ rather than x. This is always possible (just think of translating a graph to the right a units) and will turn out to be useful. It now turns out that the requirement that P and f share the same values of all their derivatives at a is equivalent to

$$a_i = f^{(n)}(a)/n!$$

(If we had expressed the polynomial as a sum of powers of x we would not have gotten such nice formulas for the coefficients.) So really, the polynomial is

$$P(x) = f(a) + f'(a)(x - a) + \frac{f''(a)}{2!}(x - a)^2 + \cdots + \frac{f^{(n)}(a)}{n!}(x - a)^n$$

This polynomial is known as the *nth degree Taylor polynomial for f at a*.

Let's see how well this polynomial does in actually approximating a function f. Remember the successive derivatives of $\tan x$ that we compute in Example 7.4.4? Let's use these to find the fifth degree Taylor polynomial for $\tan x$ near $a = 0$. The result is in Example 7.6.1.

In the first line, we compute the fifth degree Taylor polynomial of $\tan x$ about zero and name it **poly**. In the next line, we graph both functions. Let's see how the first line of code works. Basically we have used the **Sum** function to produce a sum. This

---

Example 7.6.1

```
In[52]:= (* finding the 5th degree Taylor polynomial
          for tan[x] near zero *)
       poly = Sum[
          (D[Tan[x], {x, n}] /. x → 0) x^n / n!, {n, 0, 5}
       ]
       Plot[
          {Tan[x], poly}, {x, -π / 2, π / 2},
          PlotStyle → {Thickness[.01], Thickness[.005]}
       ]
```

$$Out[52]= x + \frac{x^3}{3} + \frac{2 x^5}{15}$$

Example 7.6.1 (Continued)

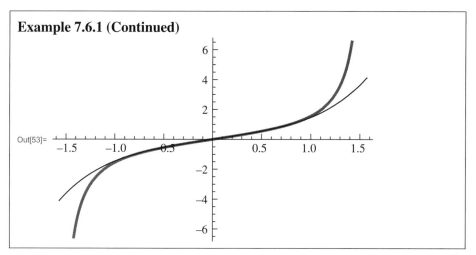

function has the same syntax as **Table** but instead of producing a list as **Table** would, it adds all the elements together. To produce the nth summand, we compute the nth derivative of $\tan x$, evaluate it at zero, and then multiply it times $x^n/n!$ We get $x + x^3/3 + 2x^5/15$ and the graph of this function closely approximates $\tan x$ near zero.

As you might have guessed, *Mathematica* has the capability to compute Taylor polynomials built in! The function **Series** will do the trick. In Example 7.6.2, we use **Series** to compute the 7th degree Taylor polynomial of $\tan x$ about zero.

Example 7.6.2

In[88]:= (* using Series and Normal to find a Taylor
         polynomial *)
      Series[Tan[x], {x, 0, 7}]
      Normal[Series[Tan[x], {x, 0, 7}]]

Out[88]= $x + \dfrac{x^3}{3} + \dfrac{2\,x^5}{15} + \dfrac{17\,x^7}{315} + O[x]^8$

Out[89]= $x + \dfrac{x^3}{3} + \dfrac{2\,x^5}{15} + \dfrac{17\,x^7}{315}$

The Taylor *series* associated to a function f is the infinite sum

$$\sum_{i=0}^{\infty} \frac{f^{(i)}(a)}{i!}(x-a)^i$$

If we take only the first n terms of the series, we obtain the nth degree Taylor polynomial. Since the series is an infinite sum, there are delicate questions surrounding its *convergence*, that is, whether or not for a specific value of x it actually adds up to a specific number or not.[5] We won't go into this important issue here. The important piece for us is that **Series** will give the Taylor polynomial up to a given degree and then include a final "term" of the form $\mathbf{O[x]^k}$ to indicate the rest of the series.[6] If we only want the Taylor polynomial, we want to ignore the "O" term and **Normal** does this for us. The **Normal** function is a specialized function that converts a few "special forms" (with series being one of them) to a "normal" expression.

We close this section with one more example: we plot the first 20 Taylor polynomials for $\sin x$ near $a = 5\pi/2$. We use **Normal** and **Series** to find the Taylor polynomials as elements of a list created with **Table**. We call this list **polys**. Then we use **Plot** to plot everything.

Example 7.6.3

```
In[54]:= (* Taylor polys for sin[x] near 5Pi/2 *)
    polys =
      Table[Normal[Series[Sin[x], {x, 5 π / 2, k}]],
        {k, 0, 20}
      ];
    Plot[{Sin[x], polys}, {x, 0, 5 π},
      PlotRange → {{0, 5 π}, {-2, 2}}
    ]
```

Out[55]=

---

[5] As strange as it sounds, you were probably introduced to infinite sums in the 4th or 5th grade! (Although your teacher didn't tell you!) Remember when you learned that $\frac{1}{3} = .333333\ldots$ with the decimal extending forever with an infinite number of 3s? This is actually the infinite sum $\frac{3}{10} + \frac{3}{100} + \frac{3}{1000} + \ldots$ and amazingly, it adds up to (or *converges* to) exactly $\frac{1}{3}$.

[6] The "O" stands for "order" and this "Big-O" notation is standard fare for representing unknown or unnamed terms in an expression that have a given order, or degree.

From the plot we see that the 20th polynomial stays close to $\sin x$ for some time. It is really pretty amazing that only measuring properties of $\sin x$ at a single point ($a = 5\pi/2$) is sufficient to build a polynomial that can so closely approximate $\sin x$ at points far from $5\pi/2$.

7.7 Integration

The second pillar of calculus is integration, which you may recall is defined via *Riemann sums*. Remember that the Riemann sum of a function $f(x)$ on an interval (a, b) is defined as follows. First, the interval is subdivided into a finite number of subintervals. These subintervals do not all need to be the same length, although in practice it is simplest to adopt this procedure and that is what we will do from now on. If there are n subintervals then each has length $(b - a)/n$ and we will call this quantity Δx (read "delta x"). Next, we "sample" the function at a point in each subinterval. This means that we pick a point in each of the subintervals and compute the value of the function at that point. The sampling set does not have to be chosen in any particular way. We could even choose randomly in each subinterval. But again, there are a few standard practices: we could always sample at the left endpoint of each subinterval, or at the midpoint, or at the right endpoint, for example. Finally, all the sampled values are added together and the sum is multiplied times Δx. This is called a *Reimann sum*. There are infinitely many different ways to form the sum, so what you get definitely depends on how the subintervals and sampling points are chosen. The big idea now, however, is to try and take the limit of what you get as the number of subintervals goes to infinity and their individual lengths go to zero. For "nice" functions this limit will exist and it is called the *integral* of $f(x)$ from a to b and is denoted by

$$\int_a^b f(x)\, dx$$

Reimann sums have a very nice graphical interpretation. Rather than first adding all the sampled values of the function and then multiplying by Δx, we could use the distributive law to first multiply each sampled value by Δx and then add the results. Each summand can now be thought of as representing the area of a rectangle whose base is the subinterval and whose height is the value of the function at the sampling point. Thus the Riemann sum represents an area and clearly this area is close to the area of the region "under" the graph of $f(x)$ if the number of rectangles is large.

We have included a rather lengthy *Mathematica* cell in the next example that uses **Manipulate** and a few other interesting functions to illustrate Riemann sums. The point of this example is not to accurately compute Riemann sums—*Mathematica* has built-in functions for that. But rather we want to show how to build fairly sophisticated *Mathematica* routines out of the basic functions that we are learning about. Let's take a look at the example and see how it works. As usual, you need to type this in and run it yourself. The **Manipulate** feature has to be seen in action to appreciate how it works.

Example 7.7.1 contains the output cell generated by the input cell in Example 7.7.2. Let's begin by looking at the output which shows the graph of a function together with a collection of rectangles beneath the curve that represent a Riemann sum. We can vary the number of subintervals in the Riemann sum by using the slider. We can also choose the sampling method, with left endpoint, midpoint, or right endpoint sampling available choices as well as a fourth choice, "Trapezoid," that we will discuss in a moment. With each choice of sampling method, the Riemann sum is computed and displayed.

The *Trapezoid rule* is closely related to the Riemann sums that have been described so far. Instead of adding the areas of rectangles that are built on the subintervals, we use trapezoids. Choose "Trapezoid" for the sampling method as well as a fairly small number of subintervals and it should be clear what we mean. It appears as though the sum of the areas of the trapezoids should do a better job (with the same number of subintervals) of approximating the area under the graph of $f(x)$ than any of the other three sampling methods, and this is the motivation behind the rule. Notice that the sum obtained using the trapezoid rule is just the average of the left and right-handed Riemann sums! This is because the trapezoid lies "halfway" between the two rectangles.

Let's examine the *Mathematica* code to see how it works. The first line defines the function that we want to integrate and the interval (a, b). Try changing these and rerunning the example. The rest of the cell is a giant **Manipulate** instruction with two controllers: n and *type*. The number of rectangles is controlled by n and the sampling method by *type*.

Look at the code for the controllers and understand it before moving your focus to the body of the **Manipulate** function. Notice that we let n range from 1 to 100 in steps of 1. It doesn't really make sense to allow fractional parts of rectangles. We initialize n to 10 and also label the slider with "No. of Rectangles." We have also added an option: **Appearance** → "**Labeled**" which causes the value of the controller to be printed to the right of the slider bar so that we do not have to "open" the controller to obtain this information. The second controller, *type*, can only take on a set of four values, so instead of displaying a slider, *Mathematica* offers up the four choices as buttons. If you haven't already checked out the tutorial on manipulate contained in the Help Files you should do so. There are lots of options for the controllers.

There are four instructions inside the body of **Manipulate**. The first computes Δx. The second computes four numbers called $w1$, $w2$, $w3$, and $w4$ that will be

Example 7.7.1

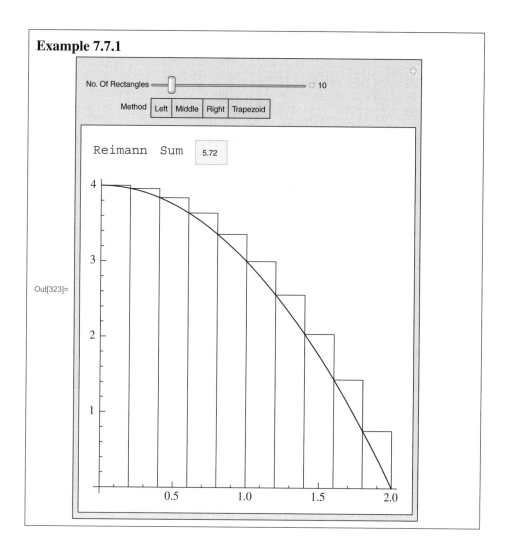

used to draw the rectangles or trapezoids depending on the choice of method. We'll come back to these in a moment. The third instruction computes the Riemann sum, and finally the fourth instruction lays out the plot and the value of the sum by using **Grid**.

Let's start by seeing how the sum is computed. Of course, it depends on the choice of sampling method, hence we have used the **Switch** function. This function will return different answers depending on different cases. The first argument to **Switch** is an expression, in this case **type**. The remaining arguments now come in pairs. If **type** matches the first argument of a pair then the second argument of that pair is returned. We have four pairs, one for each of the possible value of

Example 7.7.2

```
(* Riemann Sum demonstration *)
(* define function and interval *)
f[x_] := 4 - x^2; a = 0; b = 2 ;
Manipulate[
 (* define width of each rectangle *)
 Δx = N[(b - a) / n];
 (* get w's used for drawing rectangles *)
 {w1, w2, w3, w4} = Switch[type,
   "Left", {1, 0, 1, 0},
   "Middle", {.5, .5, .5, .5},
   "Right", {0, 1, 0, 1},
   "Trapezoid", {0, 1, 1, 0}];
 (* compute Reimann sum according to method *)
 sum = Switch[type,
   "Left", Sum[f[a + i Δx], {i, 0, n - 1}] Δx,
   "Middle", Sum[f[a + Δx / 2 + i Δx] , {i, 0, n - 1}] Δx,
   "Right",  Sum[f[a + i Δx] , {i, 1, n}] Δx,
   "Trapezoid",
   Sum[(f[a + i Δx] + f[a + (i + 1) Δx]) / 2, {i, 0, n - 1}] Δx;
 (* output layout uses nested grids *)
 Grid[{{Grid[{{"Reimann Sum", Panel[sum]}}]},
   {Plot[f[x], {x, a, b},
     AspectRatio → 1, ImageSize → {400, 400},
     (* use epilog to draw rectangles *)
     Epilog →
      Table[Line[{
         {L = a + (i - 1) Δx, 0},
         {R = a + i Δx, 0}, {R, w1 f[L] + w2 f[R]},
         {L, w3 f[L] + w4 f[R]},
         {L, 0}}], {i, 1, n}]]
   }},
  (* grid options *)
  Frame → None, Spacings → {1, 1}, Alignment → Left
 ],
 (* manipulate controllers *)
 {{n, 10, "No. Of Rectangles"}, 1, 100, 1,
  Appearance → "Labeled"},
 {{type, "Left", "Method"},
  {"Left", "Middle", "Right", "Trapezoid"}}
]
```

type. So, for example, if **type** is "Left," then **Switch** will return the left Riemann sum. Notice how this sum is computed. We use the **Sum** function to add up all the sampled values of the function. This sum is then multiplied times Δx. Notice how the sampling points are described as a plus multiples of Δx. Notice that the difference between the Left and Right Riemann sum lies with the upper and lower limits of summation: 0 and $n-1$ versus 1 and n, respectively.

After the Riemann sum is computed, we use **Plot** to draw the graph of the function and then the rectangles or trapezoids. We use three options with **Plot**: **AspectRatio**, **ImageSize**, and **Epilog**. The first two are used just to make the image look good, but **Epilog** is used to add in all the rectangles, or trapezoids, after the graph is plotted. Recall that **Epilog** is used to add graphical elements to a plot. In this case, we use **Table** to generate a list of the rectangles, or trapezoids, that are then added with the **Epilog** option. If we had only been drawing rectangles we could have used the **Rectangle** function. But since trapezoids are also a possibility, we simply use the **Line** function to draw the perimeter of each quadrilateral. Each quadrilateral has the same two vertices on the x-axis, but the upper two vertices depend on whether we are drawing a rectangle or a trapezoid. This is where the $w's$ come in. Note that **L** and **R**, which stand for Left and Right, are defined as the x-coordinates of the two lower vertices of each quadrilateral. The upper two points, which are the third and fourth vertices used in the **Line** function then have y-coordinates that are *weighted averages* of **f[L]** and **f[R]**. If you take a careful look at each of the four cases you'll discover that the *weights* **w1**, **w2**, **w3**, and **w4** are just what is needed to draw the correct shape.

Finally, the results (the plot and the value of the Riemann sum) are layed out using nested **Grid** functions. The outermost **Grid** function lays down two rows. The top row displays the value of the Riemann sum, and the bottom row displays the plot. But the top row is itself a **Grid** of two items arranged in a single row: the text "Riemann sum," and the value of the sum. Finally, the sum is displayed in a **Panel** just to make it look good. Check out the Help Files to read about **Grid**, **Panel** and the options **Frame**, **Spacings**, and **Alignment** that are all used here.

Now that we really understand what a Riemann sum is, the good news is that *Mathematica* will compute Riemann sums with the function **NIntegrate**. If you played around with Example 7.7.2, you must have found that $\int_0^2 4 - x^2 \, dx$ is about 5.33333. Example 7.7.3 computes this integral using **NIntegrate**.

Example 7.7.3

```
(* using NIntegrate to compute a Riemann sum *)
NIntegrate[4 - x^2, {x, 0, 2}]
```
Out[7]= 5.33333

The syntax for **NIntegrate** is pretty straight forward. The first argument is the integrand, or function that we want to integrate. Next we have a list declaring the variable of integration as well as the lower and upper limits of integration.

Here are a couple more examples of integrals computed using **NIntegrate**.

Example 7.7.4

In[56]:= (* a handful of integrals *)
```
NIntegrate[x³ - 3 x² + x - 6, {x, -1, 3}]
NIntegrate[Tan[x]², {x, 0, π / 4}]
NIntegrate[ArcSin[3 x] / (x² + 1), {x, 0, .3}]
NIntegrate[Exp[-x²], {x, 0, 1}]
NIntegrate[1 / x², {x, 1, Infinity}]
```

Out[56]= -28.

Out[57]= 0.214602

Out[58]= 0.1414

Out[59]= 0.746824

Out[60]= 1.

Notice that *Mathematica* can even compute improper integrals with **NIntegrate**. Recall that an integral is called *improper* if either the interval over which we wish to integrate is infinite, or the function approaches infinity somewhere on the interval of integration. In our last example, we are asking for all the area under the curve $1/x^2$ to the right of $x = 1$. This is an unbounded region and yet it has a finite area of one unit.

We'll include more examples, including multiple integrals, after we discuss antidifferentiation in the next section.

7.8 Antidifferentiation

We have seen already that given a function $f(x)$ we can create a new function $f'(x)$ by taking its derivative. If we run this process backward, it is called *antidifferentiation*. The goal now is to find a function with a given derivative. The Fundamental Theorem of Calculus, which as the name, implies must be pretty significant, relates antidifferentiation to integration in an extremely important way. If $f(x)$ is a function with antiderivative $F(x)$ (so that $F'(x) = f(x)$), then the Fundamental

Theorem of Calculus states that

$$\int_a^b f(x)\,dx = F(b) - F(a)$$

This is really a pretty amazing theorem. On the left-hand side, we have the integral of $f(x)$ over the interval $[a, b]$ which remember is defined as the limit of all possible Riemann sums as the subdivision of the interval $[a, b]$ grows finer and finer. On the right-hand side, we have the difference of the values of the antiderivative of $f(x)$ at the endpoints of the interval. At first glance, it seems totally unlikely that these things would be related at all!

The Fundamental Theorem is quite useful because it allows us to trade integration (i.e., computing Riemann sums, and perhaps wondering how accurate our estimate for the integral is) with antidifferentiation. On the other hand, before you think this is the death knell for integration, be warned that lots of functions (perhaps most functions, depending on how you count things) don't have antiderivatives! We can't prove this here, but, for example, there isn't any function whose derivative is e^{-x^2}.[7] Having just said this though, it is true that for lots and lots of simple functions we can find explicit antiderivatives. Indeed a good portion of any college calculus course is spent learning "methods of integration," which in fact are not really integration methods at all but are really antidifferentiation methods.

Happily, *Mathematica* can antidifferentiate for us! The function **Integrate** does just what we want.[8] Example 7.8.1 illustrates the use of **Integrate**.

Example 7.8.1

```
In[91]:= (* finding indefinite and definite integrals *)
     Integrate[4 - x^2, x]
     Integrate[4 - x^2, {x, 0, 2}]
```

$$Out[91]= 4\,x - \frac{x^3}{3}$$

$$Out[92]= \frac{16}{3}$$

---

[7] This isn't quite true. An equivalent form of the Fundamental Theorem asserts that $\int_0^x e^{-t^2}\,dt$ is an antiderivative of e^{-x^2}. But this antiderivative will be of no real help in evaluating integrals of e^{-x^2}.

[8] Since **Integrate** finds antiderivatives, a better name for the function would be **AntiDerivative**. But it is too late now to change the longstanding use of "integration" in place of "antidifferentiation."

We see in this example that **Integrate** can be used in two fundamentally different ways. In the first usage, we are finding an antiderivative of $4 - x^2$. This is also known as the *indefinite integral* of $4 - x^2$ and is often written as $\int 4 - x^2 \, dx$. It is easy to compute the derivative of $4x - x^3/3$ and see that we get $4 - x^2$. We could even ask *Mathematica* to do this using the differentiation function **D**.

Recall that by adding a constant to one antiderivative we obtain another antiderivative and that any two antiderivatives are related in this way. So, if we want to think about *all* antiderivatives of $4 - x^2$ we need to add an arbitrary *constant of integration* to the single answer *Mathematica* gave us, obtaining $4x - x^3/3 + C$, where C is any constant.

The second usage of **Integrate** in this example computes the *definite integral* $\int_0^2 4 - x^2 \, dx$. In this case, it computes the integral by first computing the antiderivative and then using the Fundamental Theorem. Thus we get the exact answer of $16/3$. Notice that in Example 7.7.3 the use of **NIntegrate** with this very same integral gave us the approximate answer of 5.33333. If you haven't guessed by now, the **N** in **NIntegrate** stands for numerical. Since **Integrate** will give exact answers, why should we ever use **NIntegrate**? The answer is that for simple integrals we may never need **NIntegrate**, but if we run into more complicated integrals, **Integrate** might not work for us.

Let's look at some more examples. Suppose we want to find the area trapped between the parabola $y = 9 - x^2$ and the line $y = 1 + 2x/3$. In the next Example we plot the region, find the endpoints of the interval over which we will integrate, and perform the integration.

We use **Plot** to graph both the parabola and the line, but have added the **Filling** option to shade in the region. There are several ways to use **Filling**. For example, when we only plot the graph of a single function the option **Filling** $\rightarrow$ **Axis** will fill in between the graph and the horizontal axis. The option value **Axis** can be replaced with **None, Bottom**, or **Top**. Try these out to see how they work. But if we plot multiple graphs, we can fill in between any pair. For example, if we plot three functions at once, the option **Filling** $\rightarrow$ $\{1 \rightarrow \{3\}\}$ will fill in between the first and third function. In Example 7.8.2, if we had used **Filling** $\rightarrow$ $\{1 \rightarrow \{2\}\}$ (or **Filling** $\rightarrow$ $\{2 \rightarrow \{1\}\}$) the area between the two graphs would have been filled in. However, we only want to fill above the line and below the parabola, so we have used the form **Filling** $\rightarrow \{1 \rightarrow \{\{2\}, \{\text{None, GrayLevel[.7]}\}\}\}$. With this usage, the two filling styles, **None** and **GrayLevel[.7]** apply to the regions that are below and above the second function, respectively. Exactly what we want!

After plotting the region, we use **Solve**[9] to find the horizontal endpoints of the region. Finally, we use **Integrate** to compute the definite integral. The area of this region is exactly $292\sqrt{73}/81$.

---

[9]See Chap. 8 for information on using **Solve**.

Example 7.8.2

In[61]:= (* find the shaded area *)
        (* plot the region *)

        Plot$\left[\left\{9 - x^2,\ 1 + \dfrac{2\,x}{3}\right\},\ \{x,\ -5,\ 4\},\right.$

           Filling → {1 → {{2}, {None, GrayLevel[.7]}}}
           $\Big]$

        (* find the endpoints *)
        {a, b} = x /. Solve[9 - x^2 == 1 + 2 x / 3, x]
        (* compute the integral *)
        Integrate[9 - x^2 - (1 + 2 x / 3), {x, a, b}]

Out[61]=

Out[62]= $\left\{\dfrac{1}{3}\left(-1 - \sqrt{73}\,\right),\ \dfrac{1}{3}\left(-1 + \sqrt{73}\,\right)\right\}$

Out[63]= $\dfrac{292\,\sqrt{73}}{81}$

To close out this section, suppose we want to compute the antiderivative $\int e^{-x^2}\,dx$. Look what happens if we try to use **Integrate**.

Example 7.8.3

In[64]:= (* trying to antidifferentiate e^{-x^2} *)

        Integrate$\left[\text{Exp}\left[-x^2\right],\ x\right]$

Out[64]= $\dfrac{1}{2}\,\sqrt{\pi}\ \text{Erf}[x]$

Example 7.8.3 (Continued)
> `? Erf`

Erf[z] gives the error function erf(z).

Erf[z_0, z_1] gives the generalized error function erf(z_1) – erf (z_0). ≫

If you have never seen the function **Erf[x]** before you will be wondering what is going on! Using ?**Erf** tells us that **Erf** is the error function. Still not much help, although now we know the name of the function! If we follow the link into the Help Files we discover that the *definition* of this function is

$$\text{erf}(x) = \frac{2}{\sqrt{\pi}} \int_0^x e^{-t^2}\, dt$$

So what we have just found out is that

$$\int e^{-x^2}\, dx = \frac{1}{2}\sqrt{\pi}\,\text{erf}(x) = \int_0^x e^{-t^2}\, dt$$

which is just a statement of the Fundamental Theorem of Calculus. We really aren't getting anywhere. If we wanted to use this antiderivative to compute the definite integral $\int_0^1 e^{-x^2}\, dx$ all we would be finding out is that

$$\int_0^1 e^{-x^2}\, dx = \int_0^1 e^{-t^2}\, dt$$

This is where we really need to use **NIntegrate**. The next example gives us a numerical value of the definite integral. Notice the somewhat tricky way of shading the desired region. We've created a second function that agrees with e^{-x^2} outside of the interval (0, 1), while inside (0, 1) it is zero. We can now use **Filling** to fill in between the original function, e^{-x^2}, and this new function.

Example 7.8.4
```
In[65]:= (* finding the area of the shaded region *)
         (* define lower edge of region so we can
         use Filling to shade the region *)
```

Example 7.8.4 (Continued)

```
f[x_] := If[x > 0 && x < 1, 0, Exp[-x²]];
Plot[{Exp[-x²], f[x]}, {x, -3, 3},
 Filling → {1 → {2}}
]
NIntegrate[Exp[-x^2], {x, 0, 1}]
```

Out[65]=

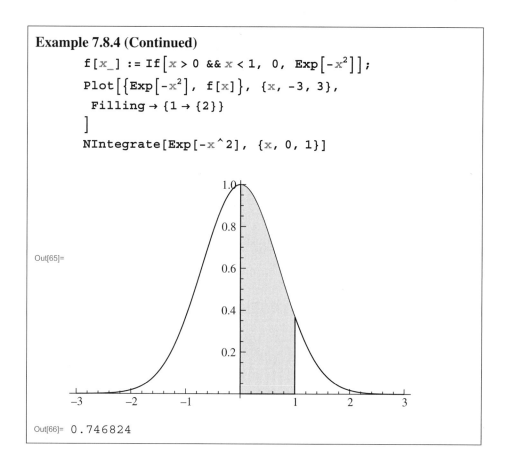

Out[66]= 0.746824

7.9 Applications of Integration

Now that we know how to use **Integrate** and **NIntegrate**, let's close out this chapter by doing a few examples.

Problem 1: Find the total length of the Lissajous knot parameterized by

$$x(t) = \cos(2t)$$
$$y(t) = \cos(3t + \pi/4)$$
$$z(t) = \cos(5t + 2).$$

The solution is displayed in Example 7.9.1. We begin by defining the curve parametrically as a *position vector*, that is, a list of its coordinate functions. Next, just for fun, we plot the curve. The **Tube** plot style gives a nice effect. In order to

Example 7.9.1

In[61]:= (* finding the length of a Lissajous
knot *)
(* define the knot curve *)
k[t_] := {Cos[2 t], Cos[3 t + Pi / 4],
Cos[5 t + 2]};
(* plot the knot *)
ParametricPlot3D[k[t], {t, 0, 2 Pi},
PlotStyle → Tube[.03]]
(* define the velocity vector *)
velocity[t_] = k'[t]
(* the speed is the length of the
velocity vector *)
speed[t] = Norm[velocity[t]]
(* integrate the speed to get the length *)
NIntegrate[speed[t], {t, 0, 2 Pi}]

Out[62]=

Out[63]= $\left\{ -2\,\text{Sin}[2\,t],\ -3\,\text{Sin}\left[\dfrac{\pi}{4}+3\,t\right],\ -5\,\text{Sin}[2+5\,t]\right\}$

Out[64]= $\sqrt{\left(4\,\text{Abs}[\text{Sin}[2\,t]]^2 + \right.}$

$\overline{\left. 9\,\text{Abs}\left[\text{Sin}\left[\dfrac{\pi}{4}+3\,t\right]\right]^2 + 25\,\text{Abs}[\text{Sin}[2+5\,t]]^2\right)}$

Out[65]= 26.4097

find the length of any parameterized curve, we need to integrate the speed, which is the length of the velocity vector. So our next step is to differentiate the position vector, which we do using the "prime" notation for derivative.

Notice that if we differentiate a list of functions, *Mathematica* simply returns the list of the derivatives of each function. This is perfect for finding the velocity vector! Finally, we use **NIntegrate** to find the length of the curve. Notice that we use the **Norm** function to find the length of the velocity vector.

Problem 2: Find the mass of a thin disk of radius 1 whose density r units from the center is given by $\delta = \sqrt{1/4 + r^2}$.

Let's introduce Cartesian coordinates with the disk centered at the origin. The density at the point (x, y) is now given by $\delta(x, y) = \sqrt{1/4 + x^2 + y^2}$ so that the disk is denser out by its edge and less dense near the center. We can calculate the mass as a double integral—we multiply each element of area by its density to get an element of mass and then sum these over the whole disk. Because of the symmetry of the disk, we may compute the mass only in the first quadrant and then multiply by four. We get the following double integral:

$$4 \int_0^1 \int_0^{\sqrt{1-x^2}} \sqrt{1/4 + x^2 + y^2} \, dy \, dx.$$

Now that we have set up the integral, which is often the hard part, it is a snap to compute this with *Mathematica*. Example 7.9.2 gives the result. Notice that when performing multiple integrals with **Integrate** we simply follow the integrand (the function to be integrated) with the limits of integration for each of the variables. Furthermore, the variables are listed in the opposite order of integration. So here we have listed the x variable first because we integrate with respect to x last.

Example 7.9.2

```
In[67]:= (* finding the mass of a thin unit disk
          whose density if Sqrt[1/4+x²+y²] *)

       4 Integrate[

           Sqrt[ 1/4 + x² + y² ], {x, 0, 1}, {y, 0, Sqrt[1 - x²]}]

        ]

Out[67]= 1/12 (-1 + 5 √5) π
```

Probem 3: Find the area of one "leaf" of the curve given in polar coordinates as

$$r(\theta) = \sin(3\theta)$$

The curve is pictured in Example 2.10.4. Remember that in polar coordinates the element of area is given by

$$dA = r\, dr\, d\theta$$

In order to find the area, we need to integrate dA over the region. Example 7.9.3 does this.

Example 7.9.3

In[68]:= (* finding the area in one leaf of r=Sin[3θ] *)
   Integrate[r, {θ, 0, π/3}, {r, 0, Sin[3 θ]}]

Out[68]= $\dfrac{\pi}{12}$

As θ goes from 0 to $\pi/3$, 3θ goes from 0 to π, and hence $\sin(3\theta)$ goes from 0 out to 1 and then back to 0. Thus we trace out one leaf of the curve in the interval $[0, \pi/3]$. So we first integrate with respect to r from the origin out to the curve, and then with respect to θ from zero to $\pi/3$.

Problem 4: Find the center of mass of the homogeneous "ice cream cone" consisting of a right circular cylinder of height 10 cm and base radius 2 cm, topped with a hemispherical scoop of ice cream.

Just for fun, let's draw a picture of the ice cream cone. We do this in Example 7.9.4, where we use **RevolutionPlot3D** to plot the cone and the scoop separately and then combine the plots using **Show**. Note that for both the cone and the scoop we describe the generating curve parametrically.

To make the problem simpler, we'll assume that the cone and scoop are a solid object of uniform density, and moreover that the density is equal to 1. By the symmetry of the object, the center of mass clearly lies on the axis of symmetry. So we only need to compute the z-coordinate of the center of mass. To do this, we need to divide the moment with respect to the xy-plane by the mass.

It is easiest to work with cylindrical coordinates. Recall that the element of volume in cylindrical coordinates is $r\, dr\, d\theta\, dz$ and since the density is 1, this is also the element of mass dm.

We can easily find the total mass by knowing the formulas for the volume of a cone and the volume of a ball. For a cone, the volume is one-third the area of the base times the height, and for a ball the volume is four-thirds π times the cube of the radius. Rather than working with the specific dimensions of the cone given in the problem, let's work more generally and assume the radius of the base of the cone is R and the height of the cone is H. Example 7.9.5 starts by computing the mass of the ice cream cone as a function of R and H.

Example 7.9.4

```
In[69]:= (* an ice cream cone *)
    (* the cone *)
    H = 5; R = 2;
    plot1 = RevolutionPlot3D[{t, H t / R}, {t, 0, R}];
    (* the scoop of ice cream *)
    plot2 = RevolutionPlot3D[
        {R Cos[t], H + R Sin[t]}, {t, 0, π / 2}
      ];
    (* combining the plots *)
    Show[plot1, plot2,
      ViewPoint → {2, 3, 3 / 4},
      PlotRange → {{-R, R}, {-R, R}, {0, H + R}}
    ]
```

Out[71]=

Next we need to find the moment of the ice cream cone with respect to the xy-plane. We can find the moment of the cone and scoop separately and add the results. The next part of Example 7.9.5 finds the moment of the cone. To do this, we must integrate $z\,dm$ over the solid.

Example 7.9.5

In[81]:= `(* finding center of mass of ice cream cone *)`
`(* mass of cone plus scoop *)`
`mass = (π R² ) H / 3 + 4 / 3 π R³ / 2`
`(* moment of cone wrt to xy-plane *)`
`momentCone = Integrate[`
   `z r, {z, 0, H}, {Θ, 0, 2 π}, {r, 0, R z / H}`
  `]`
`(* moment of scoop wrt to xy-plane *)`

`momentScoop = Integrate[`

  `z r,`

  `{z, H, H + R}, {Θ, 0, 2 π}, {r, 0, √(R² - (z - H)²)}`

  `]`

`(* height of center of mass *)`
`zbar = Simplify[(momentCone + momentScoop) / mass]`

Out[81]= $\dfrac{1}{3} H \pi R^2 + \dfrac{2 \pi R^3}{3}$

Out[82]= $\dfrac{1}{4} H^2 \pi R^2$

Out[83]= $\dfrac{1}{12} \pi R^3 (8 H + 3 R)$

Out[84]= $\dfrac{3 H^2 + 8 H R + 3 R^2}{4 H + 8 R}$

The order of integration is: r, then θ, then z. We first let r go from zero out to the wall of the cone, then sweep this radial piece once around in the θ direction to get the moment of a horizontal disk cross section, and finally integrate in the z-direction to sum the moments of all the disks. A similar calculation is then done for the moment of the scoop.

Finally, we add the two moments and divide by the mass to get the height of the center of mass as a function of both R and H. You can check that if we now substitute

$R = 2$ and $H = 5$, we get that the center of mass is at the point $(0, 0, 167/36)$ or about $(0, 0, 4.639)$. Notice that in this case the center of mass lies inside the cone. But if we were to increase R relative to H more of the mass would lie in the scoop and we would expect the center of mass to shift into the interior of the scoop. One of the Quiz questions asks you to find out what cone angle produces a center of mass that lies in the base of the cone.

7.10 Find Out More

We have tried to touch on all the basic *Mathematica* functions that can be used in typical calculus problems. But, of course, there are still a lot of *Mathematica* functions we have not mentioned. The following guides and tutorials are a good place to begin your exploration of the Help Files. As usual, these will provide more leads that can take you deeper into the documentation.

- guide/Calculus
- tutorial/ConstrainedOptimizationOverview
- tutorial/MinimizationAndMaximization
- tutorial/SumsAndProducts
- tutorial/SummationOfSeries

Constrained optimization is a huge area of mathematics that we have barely touched upon. The tutorial mentioned above, ConstrainedOptimizationOverview, is quite comprehensive and will lead you in quite a few directions, including the important area of *Linear Programming*.

While we mentioned Taylor polynomials and the useful **Series** function, we did not really discuss infinite sums. *Mathematica* has a number of functions that can help you investigate these to determine if they are convergent or not, and if so, what they add up to. The last two tutorials mentioned above are a good place to start.

Quiz

1. Find the limit of $\frac{\sin x^2}{x}$ as $x \to 0$.

2. Suppose $f(x) = -x^3 + x + 1$. Compute $f'(1)$ and determine the equation of the line tangent to the graph of $f(x)$ at the point $(1, f(1))$. Plot both $f(x)$ and the tangent line.

3. Can you repeat the last exercise, but embed it all within **Manipulate** so that moving a slider changes the point of tangency between the line and the cubic?

4. Find the closest point on the hyperbola $x^2 - y^2 = 1$ to the point $(3, 5)$.

5. What is the minimum distance between any point on the circle $(x - .5)^2 + (y - 1)^2 = .25$ and any point on the hyperbola $x^2 - y^2 = 1$?

6. Plot the circle and hyperbola of the last question and the line segment connecting the two point. (Hint: You could use **ParametricPlot** to plot the hyperbola and then **Epilog** to throw in the circle and the line segment. Note that $(\cosh t, \sinh t)$ parameterizes the hyperbola.)

7. Find the area of the bounded region trapped between the graphs of $y = e^x$ and $y = 4 - x^2$.

8. Suppose a thin homogenous metal plate has the shape of the region of the last question. Find its center of mass.

9. Suppose the axis of two solid right-circular cylinders, each with a radius of 1 unit, meet at a right angle. Find the volume of material that lies in both cylinders.

10. What value of R makes the formula for the z-coordinate of the center of mass found in Example 7.9.5 equal to H? (Use **Solve** to find out.) When R has this value the center of mass lies in the base of the cone. What is cone angle when this is true?

CHAPTER 8

Solving Equations

Many problems in mathematics ultimately boil down to solving an equation, or system of equations. *Mathematica* has a variety of tools for solving equations, which we'll learn about in this chapter.

8.1 Polynomial Equations

Let's start by solving a simple quadratic equation, $x^2 + 3x - 5 = 0$. Example 8.1.1 shows how to use the function **Solve** to do this.

Example 8.1.1

In[1]:= (* solving a polynomial equation *)
Solve$\left[x^2 + 3\,x - 5 == 0,\ x\right]$

Out[1]= $\left\{\left\{x \to \dfrac{1}{2}\left(-3 - \sqrt{29}\right)\right\},\ \left\{x \to \dfrac{1}{2}\left(-3 + \sqrt{29}\right)\right\}\right\}$

Solve takes two arguments. The first is the equation we want to solve and the second is the variable that we want to solve for. Notice that we have used double equal signs in the equation. Recall that a single equal sign is not used to represent an abstract equation, but rather to replace the left-hand side of the expression with the right-hand side. For example **a=2** assigns the value 2 to the variable *a*.

Mathematica finds that there are two *solutions*, that is, values of *x* that make the equation true. In general, any number that makes a polynomial zero is called a *root* of the polynomial. So in this case, the solutions of the equation are also the roots of the polynomial $x^2 + 3x - 5$. Instead of just listing the solutions, *Mathematica* gives *replacement rules* of the form $\{x \rightarrow x_0\}$, where x_0 is a solution to the equation. It is easy to get from here to a list of just the solutions if that is what we want, but it is also convenient for some purposes to have the solutions listed as replacement rules.

Example 8.1.2 illustrates using the replacement operator **/.** to place the solutions in a list, which we then name **roots**.

Example 8.1.2

In[2]:= (* creating a list of the solutions *)

roots = x /. Solve$\left[x^4 - 2\,x^3 + x + 5 == 0,\ x\right]$

Out[2]= $\left\{ \dfrac{1}{2}\left(1 - \sqrt{3 - 2\,i\,\sqrt{19}}\,\right),\ \dfrac{1}{2}\left(1 + \sqrt{3 - 2\,i\,\sqrt{19}}\,\right),\right.$

$\left. \dfrac{1}{2}\left(1 - \sqrt{3 + 2\,i\,\sqrt{19}}\,\right),\ \dfrac{1}{2}\left(1 + \sqrt{3 + 2\,i\,\sqrt{19}}\,\right)\right\}$

Notice that in each of the previous examples the number of solutions has always been equal to the degree of the polynomial. This is an extremely important fact known as the Fundamental Theorem of Algebra, that is, every polynomial of degree *n* has *n* roots. Remember though that a root might be "repeated," so that we have to count the roots with their *multiplicities* to get a total of *n* roots. For example, the 10th degree polynomial $(x - 1)^6(x + 2)^4$ has only two distinct roots, 1 and −2, but 1 is repeated 6 times, and −2 is repeated 4 times, for a total of 10.

In both of these examples, we have solved a polynomial equation of degree less than 5. *Mathematica* can always solve such an equation exactly using radicals, that is, only using the operations of addition, subtraction, multiplication, division, and taking roots. But it is a famous consequence of Galois[1] theory that this is not possible in general for polynomial equations of degree 5 or more. It might be possible in special cases, but no general formula (like the famous quadratic formula in the degree two case) can exist. So what happens if we try to use **Solve** for a higher degree equation?

---

[1]Éveriste Galois (1811–1832) is famous for having solved the problem of deciding when a polynomial equation can be solved by radicals. His work led to what is now called Galois theory, an important branch of algebra. Galois theory can be used to prove that the famous Greek straightedge-and-compass problems of trisecting the angle, squaring the circle, and doubling the cube are impossible. Tragically, Galois died in a duel at the age of 20.

In Example 8.1.3, *Mathematica* has been unable to find the solutions, or *roots*, exactly. It simply lists as the solutions the five roots **Root[f, 1]**, **Root[f, 2]**, **Root[f, 2]**, **Root[f, 4]**, and **Root[f, 5]**, where **f** is the function **5+#1-2#1^3+#1^5&**, which if you look closely, you will see is our original polynomial. (It is written here as a function rule.)

Example 8.1.3

In[3]:= (* trying to solve a quintic may not work *)

Solve$\left[x^5 - 2\,x^3 + x + 5 == 0,\ x\right]$

Out[3]= $\left\{\left\{x \to \text{Root}\left[5 + \#1 - 2\,\#1^3 + \#1^5\ \&,\ 1\right]\right\},\right.$

$\qquad \left\{x \to \text{Root}\left[5 + \#1 - 2\,\#1^3 + \#1^5\ \&,\ 2\right]\right\},$

$\qquad \left\{x \to \text{Root}\left[5 + \#1 - 2\,\#1^3 + \#1^5\ \&,\ 3\right]\right\},$

$\qquad \left\{x \to \text{Root}\left[5 + \#1 - 2\,\#1^3 + \#1^5\ \&,\ 4\right]\right\},$

$\qquad \left.\left\{x \to \text{Root}\left[5 + \#1 - 2\,\#1^3 + \#1^5\ \&,\ 5\right]\right\}\right\}$

Anyway, all of this is of little help if what we want to know are the solutions to the equation. The problem, of course, is that *Mathematica* is trying to find the roots exactly. If we only ask for approximations we can get the solutions by using the *numerical* solve function. As the following examples show, we can either call the numerical solve function directly using **NSolve**, or continue to use **Solve** but switch away from exact numbers for the coefficients.

Example 8.1.4

In[9]:= (* getting numerical answers by using NSolve,
     or Solve with approximate coefficients *)

NSolve$\left[x^5 - 2\,x^3 + x + 5 == 0,\ x\right]$

Solve$\left[x^5 - 2.0\,x^3 + x + 5 == 0,\ x\right]$

Out[9]= $\{\{x \to -1.65477\},\ \{x \to -0.527822 - 1.02701\,\mathrm{i}\},$

$\qquad \{x \to -0.527822 + 1.02701\,\mathrm{i}\},$

$\qquad \{x \to 1.35521 - 0.655415\,\mathrm{i}\},$

$\qquad \{x \to 1.35521 + 0.655415\,\mathrm{i}\}\}$

Out[10]= $\{\{x \to -1.65477\},\ \{x \to -0.527822 - 1.02701\,\mathrm{i}\},$

$\qquad \{x \to -0.527822 + 1.02701\,\mathrm{i}\},$

$\qquad \{x \to 1.35521 - 0.655415\,\mathrm{i}\},$

$\qquad \{x \to 1.35521 + 0.655415\,\mathrm{i}\}\}$

We close this section with a nice little routine to plot the roots of a polynomial in the complex plane. Let's look at the example and then see how it works.

Example 8.1.5

```
In[6]:= (* a routine to plot the roots of a
    polynomial *)
  rootPlot[poly_] := Module[{roots},
    roots = x /. NSolve[poly == 0, x];
    (* convert list of roots to list of
    points and plot them *)
    ListPlot[
    {Re[#], Im[#]} & /@ roots,
    PlotStyle → {PointSize[.02]},
    AspectRatio → Automatic,
    PlotRange → {{-2, 2}, {-2, 2}},
    Epilog → {Circle[{0, 0}, 1]}]
    ]
```

In Example 8.1.5, we have used the **Module** structure with a single local variable, **roots**, to define the function **rootPlot**. The first instruction is

$$\text{roots} = x/.\text{NSolve}[\text{poly}==0, x]$$

which uses **NSolve** to find the roots and then the substitution operator to make a list of the roots, which we name **roots**. The next instruction is **ListPlot** which is used to plot the roots and which uses a few options. Since **ListPlot** expects a list of ordered pairs of real numbers we need to first turn the list of roots (which are complex numbers) into coordinate pairs. We do this by creating a pure function, **{Re[#], Im[#]}&**, that will take a single complex number as input and return the list of its real and imaginary parts. Just what we need! Next we **Map** this (using the /@ construction) onto all the elements in **roots**, producing a list of points ready for **ListPlot**. We then throw in a couple of familiar options, **PlotStyle, AspectRatio**, and **PlotRange**, as well as **Epilog** to plot the unit circle after the roots are plotted. Plotting the unit circle is unnecessary, but will be interesting in Example 8.1.6, where we show **rootPlot** in action.

It is interesting to vary a single coefficient in a polynomial and watch how the roots change. We illustrate this in Example 8.1.6 by using **Manipulate**. Try this out, and as you play with the value of a, notice that something interesting happens when a is close to 1.75. Also, notice that when $a = 0$ the roots all lie on the unit circle.

As *a* passes though 1.75, the three pairs of roots which are lying inside the unit circle each come together and then separate again. There must be a value of *a* for which the polynomial has three repeated roots instead of all 12 roots being distinct. How can we find the value of *a* that makes this true? It is an interesting fact that a polynomial has repeated roots if and only if the polynomial and its derivative (which is itself a polynomial of one less degree) have a root in common. And there is a way to decide if two polynomials have a root in common by computing their *resultant*.

Example 8.1.6

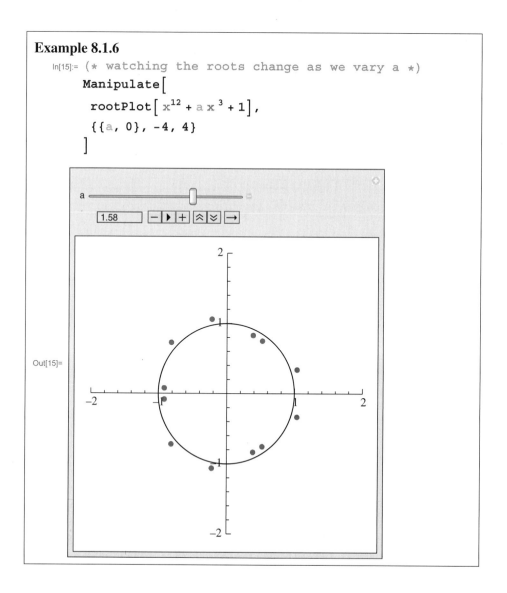

If

$$f(x) = a_n x^n + a_{n-1} x^{n-1} + \cdots + a_0$$

$$g(x) = b_m x^m + b_{m-1} x^{m-1} + \cdots + b_0$$

are polynomials, with $a_n \neq 0$ and $b_m \neq 0$, and if the roots of f are $r_1, r_2, \ldots, r_n$, then the *resultant* of f and g is defined to be

$$a_n^m g(r_1) g(r_2) \ldots g(r_n)$$

Notice that if one of the roots of f is also a root of g then the resultant will be zero. Conversely, if the resultant is zero, then one of the factors $g(r_i)$ must be zero and hence r_i is a root of g. The upshot of all this is that two polynomials have a root in common if and only if their resultant is zero. And, lucky for us, *Mathematica* will compute the resultant of two polynomials. Example 8.1.7 shows how we can use the **Resultant** function to figure out what value of a makes our polynomial have multiple roots.

Example 8.1.7

In[16]:= (* finding the value of a that makes p
         have multiple roots *)
      p = x^12 + a x^3 + 1;
      r = Resultant[p, D[p, x], x]

Out[17]= $\left(20\,736 - 2187\,a^4\right)^3$

In[18]:= Reduce[r == 0, a]

Out[18]= $a == -\dfrac{4}{3^{3/4}} \;||\; a == -\dfrac{4\,i}{3^{3/4}} \;||\; a == \dfrac{4\,i}{3^{3/4}} \;||\; a == \dfrac{4}{3^{3/4}}$

In[21]:= N$\left[\dfrac{4}{3^{3/4}}, 10\right]$

Out[21]= 1.754765351

What we have done here is first find the resultant of the polynomial and its derivative. The resultant is a function of a since the polynomial and its derivative depend on a. (If two polynomials have integer coefficients, then their resultant will be an integer. But if the coefficients of the polynomials involve parameters, then so will the resultant.) We then use **Reduce** to find out what value of a makes the resultant zero. Finally, we use the numerical function **N** to estimate one of the possibilities.

It turns out that if a is approximately ± 1.754765351 or $\pm 1.754765351i$ then the original polynomial $x^{12} + ax^3 + 1$ will have repeated roots. This is quite close to the 1.75 that we estimated by using **Manipulate**.

Reduce is a close relative of **Solve**. Its primary function is to reduce an equation, or set of equations, to a simpler form so that solutions to the original set of equations are described more explicitly. Unlike **Solve**, **Reduce** will not produce replacement rules as output because it returns a set of equations (or logical combinations of equations) equivalent to the original set. **Reduce** often produces output containing logical connectors such as and (**&&**) or or (**||**). In this case, using **Reduce** is quite nice in comparison to **Solve**. Since there are a total of 12 roots, but only four distinct roots, the list of replacement rules returned by **Solve** would have been rather redundant.

The theory of resultants is quite interesting but we will not go into it more in this book. Check out the Help Files to find out more about the **Resultant** function.

8.2 Systems of Polynomial Equations

There are many cases where we want to find a solution to several equations simultaneously. *Mathematica* can easily do this too. Suppose, for example, we want to find out where the line $y = x + 2$ intersects the parabola $y = 16 - x^2$. We can use **Solve** or **NSolve** to do this. We simply need to replace the equation that we want to solve with a list of equations. And of course, we need to replace the variable that we want to solve for with a list of variables. Example 8.2.1 shows how to do this.

Example 8.2.1

In[4]:= (* solving a system of equations *)
    Solve[{y == x + 2, y == -x^2 + 16}, {x, y}]

Out[4]= $\left\{\left\{y \to \frac{1}{2}\left(3 - \sqrt{57}\right), x \to \frac{1}{2}\left(-1 - \sqrt{57}\right)\right\},\right.$
$\left.\left\{y \to \frac{1}{2}\left(3 + \sqrt{57}\right), x \to \frac{1}{2}\left(-1 + \sqrt{57}\right)\right\}\right\}$

In[5]:= **N[%]**

Out[5]= {{y → -2.27492, x → -4.27492}, {y → 5.27492, x → 3.27492}}

Notice that immediately after finding the solutions exactly we used **N[%]** to get numerical estimates for the solutions.

Since these solutions represent the points where the line and the parabola intersect, let's graph the curves and try to read the solutions off of the graph. Of course, it will not be easy to get very accurate answers this way. But a nice feature of *Mathematica* will allow us to read off the coordinates of the mouse as we move it over the plot. This can be handy for getting a visual handle for what is going on. Consider the following example.

Example 8.2.2

In[222]:= (* displaying the mouse position dynamically *)
    Plot[{x + 2, -x^2 + 16}, {x, -5, 4}]
    Dynamic[MousePosition["Graphics"]]

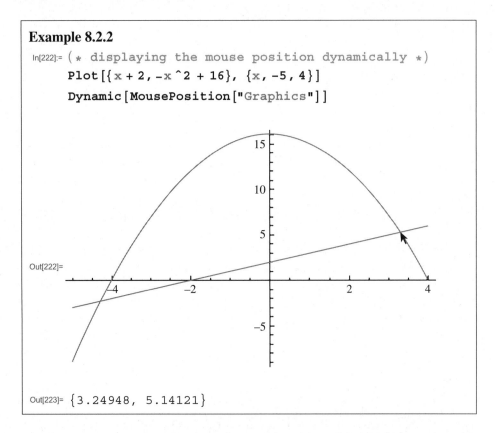

Out[223]= {3.24948, 5.14121}

Here we have plotted the line and the parabola but also included the instruction **Dynamic[MousePosition["Graphics"]]**. This statement causes the coordinates of the mouse to be displayed whenever the mouse is located inside graphical output. In this case, moving the mouse cursor to the rightmost point of intersection of the line and the parabola causes the coordinates {3.24948, 5.14121} to be displayed. As we move the mouse, these coordinates change continuously. If we move the mouse outside of the graphic, the coordinates change to **None**. Try it! Of course, we get a much more accurate answer by using **Solve** or **NSolve**. Still, plotting the functions can be quite helpful for seeing what is going on and the dynamic mouse position function is really useful. In fact, we'll see a great application of it in Chap. 10.

If we had been solving Example 8.2.3 by hand, we probably would have first eliminated one of the variables, say y, to obtain a single equation in x. We would have solved that equation for x and then substituted those solutions back into one of the original equations to solve for y. We can go through this process with *Mathematica* by using the **Eliminate** function. Let's see how to do this.

In the first cell of the Example 8.2.3, we have used **Eliminate** to eliminate the variable y. Notice that we simply list the two equations and then the variable that we want to eliminate. We could have just as easily eliminated x rather than y. We then use **NSolve** applied to the previous result to solve the single remaining equation for x. We then substitute these solutions back into the original equation $y = x + 2$. Here is an example where we took advantage of **NSolve**'s output being in the form of substitution rules to easily substitute the solutions into the original equation. In this example, we have twice referred to the previous output by using the % symbol, so keep in mind that if we rerun the second and third cells in this example (without rerunning the first cell) it will not run properly since % will not refer to the correct thing.

Example 8.2.3

In[227]:= (* using Eliminate to solve a pair of equations *)
            Eliminate[{y == x + 2, y == -x^2 + 16}, y]

Out[227]= $x + x^2 == 14$

In[228]:= (* solving for x *)
            NSolve[%, x]

Out[228]= $\{\{x \to -4.27492\}, \{x \to 3.27492\}\}$

In[229]:= (* solving for y *)
            x + 2 /. %

Out[229]= $\{-2.27492, 5.27492\}$

Eliminate can be used in much greater generality. For example, we could eliminate three variables from five equations that involve six variables. Read the Help Files for more information on **Eliminate**.

8.3 Systems of Linear Equations

Solving systems of linear equations is a major part of *linear algebra* and is worth a detour. Suppose we want to solve the system of equations

$$3x + 2y - z + w = 0$$

$$x - 3z = -1$$

$$-y + w = 2$$

Of course, linear equations are polynomial equations (of degree one) so we could just proceed as in the last section. We do this in Example 8.3.1.

Example 8.3.1

In[236]:= (* solving a system of linear equations *)
    Solve[{3 x + 2 y - z + w == 0, x - 3 z == -1, -y + w == 2},
    {x, y, z}]

Out[236]= $\left\{\left\{x \rightarrow \frac{1}{8} \ (13 - 9 \ w), \ y \rightarrow -2 + w, \ z \rightarrow \frac{1}{8} \ (7 - 3 \ w)\right\}\right\}$

In this case, notice that the system has more variables (four) than equations (three). Such a system is called *underdetermined* and it either has no solutions at all or it has infinitely many.[2] Because there are more variables than equations we have chosen to solve the system only for three of the variables in terms of the fourth. As it turns out, there are infinitely many possible solutions, one for each value of w. Notice that the substitution rules returned by **Solve** show us this since each variable is given in terms of w.

While the above example solves our problem, it is better to take the point of linear algebra and view the system of equations in terms of the following matrix equation.

$$\begin{pmatrix} 3 & 2 & -1 & 1 \\ 1 & 0 & -3 & 0 \\ 0 & -1 & 0 & 1 \end{pmatrix} \begin{pmatrix} x \\ y \\ z \\ w \end{pmatrix} = \begin{pmatrix} 0 \\ -1 \\ 2 \end{pmatrix}$$

There are very well-developed methods for solving matrix equations of the form $Ax = b$ and the *Mathematica* function **LinearSolve** is just what we want. Example 8.3.2 shows how to solve this system using **LinearSolve**. Notice that we first define the *coefficient* matrix **A** and the constant *vector* **b** and then call **LinearSolve** with the two arguments **A** and **b**.

Notice that **LinearSolve** has given us only one solution whereas **Solve** gave infinitely many! (In fact, if you look carefully, **LinearSolve** has given the single

---

[2]If there are more equations than variables, the system is called *overdetermined* and it may or may not have any solutions. If the number of equations matches the number of variables, the system is called *square*.

Example 8.3.2

In[233]:= (* solving a system of linear equations *)
      A = {{3, 2, -1, 1}, {1, 0, -3, 0}, {0, -1, 0, 1}}
      b = {0, -1, 2}
      LinearSolve[A, b]

Out[233]= {{3, 2, -1, 1}, {1, 0, -3, 0}, {0, -1, 0, 1}}

Out[234]= {0, -1, 2}

Out[235]= $\left\{ \dfrac{13}{8}, -2, \dfrac{7}{8}, 0 \right\}$

---

solution corresponding to $w = 0$.) What is going on here? The answer lies with the associated *homogeneous* system $Ax = 0$. The set of all solutions to this system is called the *null space* of A and obviously includes the zero vector since $A0 = 0$. In general, if there are multiple solutions to the original system $Ax = b$, then they will differ by something in the null space. Suppose that x_1 and x_2 solve $Ax = b$. Then

$$A(x_1 - x_2) = Ax_1 - Ax_2$$
$$= b - b$$
$$= 0$$

so we see that $x_1 - x_2$ is in the null space of A. To find all the solutions to $Ax = b$ using **LinearSolve** we need to also use the *Mathematica* function **NullSpace** to find the null space of A. We can then take the single solution given by **LinearSolve** and add to it any vector in the null space. Let's see what the null space of A is.

Example 8.3.3

In[237]:= (* finding the nullspace of the coefficient
      matrix A *)
      NullSpace[A]

Out[237]= {{-9, 8, -3, 8}}

Mathematica tells us that the vector $(-9, 8, -3, 8)$ is in the null space. But, of course, if a vector is in the null space of A, then so is any multiple of it, since $A(cx) = cAx$. So, in this case, we see that the null space of A is the line consisting of all multiples of $(-9, 8, -3, 8)$. It might turn out that a matrix A has a two or three or even higher dimensional null space in which case **NullSpace** will return a set of *linearly independent basis* vectors that *span* the null space. In this case, the null space will consist of all sums of multiples of these vectors.

Returning to our original problem, we finally see that the set of all solutions consists of

$$(13/8, -2, 7/8, 0) + w(-9, 8, -3, 8)$$

where w is any constant. Of course, this is exactly the same set of solutions that we found with **Solve**!

Let's look at one more example.

Example 8.3.4

In[246]:= (* solving a square system of equations *)
```
A = {{-5, 9, 1, 14}, {1, -1, -1, -2}, {-2, 2, 2, 4},
     {-3, 5, 1, 8}};
b = {-6, -2, 4, -2};
NullSpace[A]
LinearSolve[A, b]
```
Out[248]= {{1, -1, 0, 1}, {2, 1, 1, 0}}

Out[249]= {-6, -4, 0, 0}

In this case, the coefficient matrix is a square 4 by 4 matrix, so we are dealing with 4 equations in 4 variables. The null space of A is two dimensional, being *spanned* by the vectors $(1, -1, 0, 1)$ and $(2, 1, 1, 0)$. Notice that these two vectors are linearly independent, that is, neither is a multiple of the other, so they really do span a plane, not a line. Any linear combination of them, that is, a sum of multiples of them, is in the null space of A and will be sent to zero by A. With $b = (-6, -2, 4, -2)$, **LinearSolve** tells us that $(-6, -4, 0, 0)$ is a solution. So finally, the set of *all* solutions is the two-dimensional set of vectors

$$(-6, -2, 4, -2) + x(1, -1, 0, 1) + y(2, 1, 1, 0)$$

where x and y are any two constants.

In the previous example, we found that the equation $Ax = b$ had infinitely many solutions. But this does not have to be the case. If we change b just a little bit the situation can change a lot. Consider the following example.

Example 8.3.5

```
In[258]:= (* change b slightly and there are no solutions *)
     A = {{-5, 9, 1, 14}, {1, -1, -1, -2}, {-2, 2, 2, 4},
        {-3, 5, 1, 8}};
     b = {-6, -2, 0, -2};
     NullSpace[A]
     LinearSolve[A, b]
```

Out[260]= {{1, -1, 0, 1}, {2, 1, 1, 0}}

LinearSolve::nosol : Linear equation encountered that has no solution. ≫

Out[261]= LinearSolve[{{-5, 9, 1, 14}, {1, -1, -1, -2},
     {-2, 2, 2, 4}, {-3, 5, 1, 8}}, {-6, -2, 0, -2}]

We now see that there are no solutions at all!

Recall that associated to any square matrix is a number called its *determinant*. *Mathematica* will compute determinants as seen below.

Example 8.3.6

```
In[262]:= (* computing the determinant of the coefficient
     matrix *)
     Det[A]
```

Out[262]= 0

It is not surprising that the determinant of A is zero. In fact, a square matrix has a trivial null space consisting of only the zero vector if and only if its determinant is not zero. Since we have already seen that A has a nontrivial null space, it must be the case that the determinant of A is zero.

Mathematica has lots of functions that are useful in dealing with matrices and important problems in linear algebra. We'll close this section by describing just one more function, the **Inverse** function. Some, but not all, square matrices have an *inverse* matrix. The inverse of a matrix is entirely analogous to the (multiplicative) inverse of a number. Suppose we wanted to solve the equation

$$3x = 12$$

If we multiply both sides of the equation by 1/3 we obtain

$$x = 4$$

and the solution we are looking for. The number $1/3$ is the *multiplicative inverse* of 3. When we multiply 3 and $1/3$ together we obtain 1, which is the *multiplicative identity*. The number 1 is called this because it has the property that $1 \cdot x = x \cdot 1 = x$ for all numbers x. Wouldn't it be great if we could solve the matrix equation

$$Ax = b$$

in the same way, that is, by multiplying both sides of the equation by the "inverse" of A.

Continuing with the analogy to numbers, notice that the *identity matrix*

$$I = \begin{pmatrix} 1 & 0 & \cdots & 0 \\ 0 & 1 & \cdots & 0 \\ \vdots & \vdots & & \vdots \\ 0 & 0 & \cdots & 1 \end{pmatrix}$$

which consists of all zeroes except for 1's down the diagonal, plays the role analogous to the number 1. That it, $Ix = x$ for all vectors x. The inverse of the square matrix A is now defined to be the matrix B such that $AB = BA = I$. So, the upshot of all of this is that, if we want to solve $Ax = b$, where A is square, and IF A has an inverse, which from now on we will denote A^{-1}, then we can multiply both sides of the equation by A^{-1} to get

$$Ax = b$$
$$A^{-1}Ax = A^{-1}b$$
$$Ix = A^{-1}b$$
$$x = A^{-1}b$$

and we have solved the equation! The problem is that, not all matrices have inverses! Only those with nonzero determinant do. However, if A does have an inverse then we can find it with *Mathematica* by using the **Inverse** function. In the following example we find a few inverses and also try to find an inverse when it does not exist.

Notice that in the third example the inverse of c does not exist. *Mathematica* tells us instead that the matrix is *singular* which is another word for having zero determinant, or in other words, not having an inverse. Notice also that in the second

---

Example 8.3.7

In[263]:= (* finding the inverses of a few matrices *)
     a = {{2, 1}, {3, 4}};
     b = {{1, 2, 3}, {0, -2, 1}, {0, 2, 0}};
     c = {{1, 0}, {0, 0}};
     Inverse[a]
     MatrixForm[Inverse[b]]
     Inverse[c]

Out[266]= $\left\{\left\{\dfrac{4}{5}, -\dfrac{1}{5}\right\}, \left\{-\dfrac{3}{5}, \dfrac{2}{5}\right\}\right\}$

Out[267]//MatrixForm=
$$\begin{pmatrix} 1 & -3 & -4 \\ 0 & 0 & \frac{1}{2} \\ 0 & 1 & 1 \end{pmatrix}$$

Inverse::sing : Matrix {{1, 0}, {0, 0}} is singular. ≫

Out[268]= Inverse[{{1, 0}, {0, 0}}]

---

example we have used the formatting function **MatrixForm** in order to display the inverse of b as a matrix rather than a list of its rows.

8.4 Nonpolynomial Equations

Polynomial equations are really quite special. What if we need to solve equations like $x^2 = e^x$, $x = \cos x$, $\log x + \log 2 = 1/x$, or $\sin 2x + \tan x = 2x$? These equations involve exponential, logarithmic, and trigonometric functions, rather than just polynomials. We can still try to use **Solve** but let's see what happens.

In Example 8.4.1, notice that *Mathematica* issued us warnings with all four equations. With the first and third equations it still produced some kind of answer, but for the second and fourth it didn't come up with anything. It is not because these equations don't have any solutions! It's just that **Solve** is having a hard time. It turns out that **NSolve** doesn't do much better, although it does give numerical answers with the first and third equations rather than using the unfamiliar **ProductLog**[3] function.

---

[3]Entering **?ProductLog** we find that **ProductLog[z]**, by *definition*, is the number w that satisfies $z = we^w$. We won't have anything more to say about this somewhat esoteric function. Consult the Help Files for more information.

Example 8.4.1

In[26]:= `Solve[x^2 == Exp[x], x]`
`Solve[x == Cos[x], x]`
`Solve[Log[x] + Log[2] == 1 / x, x]`
`Solve[Sin[2 x] + Tan[x] == 2 x, x]`

InverseFunction::ifun :
Inverse functions are being used. Values may be lost for multivalued inverses. ≫

InverseFunction::ifun :
Inverse functions are being used. Values may be lost for multivalued inverses. ≫

Solve::ifun : Inverse functions are being used by Solve, so some
solutions may not be found; use Reduce for complete solution information. ≫

Out[26]= $\left\{\left\{x \to -2\ \text{ProductLog}\left[-\frac{1}{2}\right]\right\}, \left\{x \to -2\ \text{ProductLog}\left[\frac{1}{2}\right]\right\}\right\}$

Solve::tdep : The equations appear to involve
the variables to be solved for in an essentially non–algebraic way. ≫

Out[27]= `Solve[x == Cos[x], x]`

InverseFunction::ifun :
Inverse functions are being used. Values may be lost for multivalued inverses. ≫

Solve::ifun : Inverse functions are being used by Solve, so some
solutions may not be found; use Reduce for complete solution information. ≫

Out[28]= $\left\{\left\{x \to \dfrac{1}{\text{ProductLog}[2]}\right\}\right\}$

Solve::tdep : The equations appear to involve
the variables to be solved for in an essentially non–algebraic way. ≫

Out[29]= `Solve[Sin[2 x] + Tan[x] == 2 x, x]`

An alternative to **Solve** or **NSolve** that is usually quite successful is the **FindRoot** function. But to use it we need to first provide a numerical estimate of the solution. Let's consider the first equation, $x^2 = e^x$. In the next example we plot the two curves.

Example 8.4.2

In[269]:= `(* trying to find where x^2=e^x *)`
`Plot[{Exp[x], x^2}, {x, -1, 1}]`

Example 8.4.2 (Continued)

Out[269]=

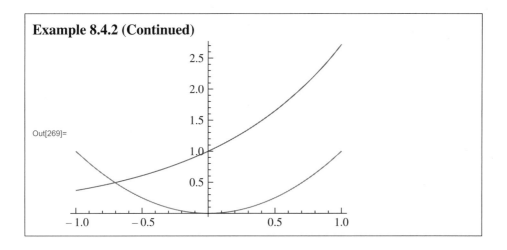

The two curves obviously intersect in exactly one point and that point is near $x = -.7$. Let's use **FindRoot** with an initial guess of $-.7$.

Example 8.4.3

```
In[270]:= (* using FindRoot to solve the equation *)
        FindRoot[x^2 == Exp[x], {x, -.7}]
```

Out[270]= $\{x \rightarrow -0.703467\}$

Notice that **FindRoot** takes two arguments. The first is the equation that we want to solve (or a list of equations to solve simultaneously) and the second is a list of the variable to solve for together with our initial guess. It turns out that for this particular equation the initial guess does not need to be that close for **FindRoot** to find the solution. But this is not always the case. Let's look at an example that shows that the outcome can be very sensitive to the initial guess.

Let's try to solve $1 + x^3/100 = x \sin x$. To get an idea of what is going on we have graphed both functions in Example 8.4.4. From the graphs we can see that there are exactly seven solutions. (There cannot be any solutions to the left or right of what we have graphed.) Looking at the plot, it is clear that the solutions lie near $-6, -3, -1, 1, 3, 7$, and 8, so let's use these as initial guesses with **FindRoot**. We do this in Example 8.4.5 and find that the solutions are approximately $\{-6.07699, -2.87281, -1.10446, 1.1244, 2.68051, 6.9627, 8.44329\}$. So far so good.

Example 8.4.4

In[123]:= (* looking for solutions to 1+x^3/100==x Sin[x] *)
        Plot[{x^3/100 + 1, x Sin[x]}, {x, -13, 13}]

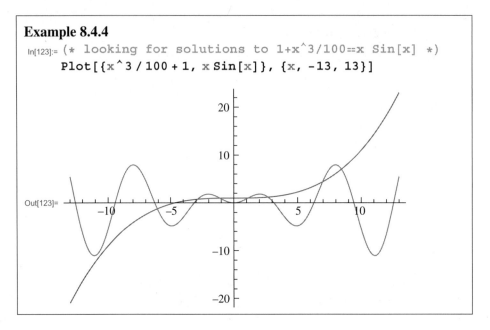

Out[123]=

In Example 8.4.5, we **Map** the pure function **FindRoot[**$1 + \frac{x^3}{100} ==$ **Sin[x]]&** onto the list of initial guesses.

Example 8.4.5

In[24]:= (* finding the solutions to 1+x^3/100==
        x Sin[x] *)

$$\text{FindRoot}\left[1 + \frac{x^3}{100} == x\,\text{Sin}[x], \{x, \#\}\right] \& /@$$

        {-6, -3, -1, 1, 3, 7, 8}

Out[24]= {{x → -6.07699}, {x → -2.87281},
        {x → -1.10446}, {x → 1.1244},
        {x → 2.68051}, {x → 6.9627}, {x → 8.44329}}

Notice in Example 8.4.6, however, that the output of **FindRoot** can vary radically with small changes in the initial guess. Here we have used four different initial guesses. The first and third differ by 0.003 and lead to the same solution, yet in between these initial values are two others that lead to different solutions! The problem is not with **FindRoot** per se, but rather is inherent in this type of problem. In fact, this type of phenomenon is usually referred to as *sensitive dependence to initial conditions* and is a hallmark of *dynamical systems* which we will discuss in Chap. 10. The lesson to be learned here is that the initial guess may not lead to where you think it will. Fortunately an initial guess that is close to a solution will often lead to that solution. A little investigation before using **FindRoot**, like our plot in Example 8.4.4, can go a long way toward guaranteeing success.

Example 8.4.6

In[25]:= (* FindRoot is sensitive to initial guess *)

$$\text{FindRoot}\left[1 + \frac{x^3}{100} == x \, \text{Sin}[x], \{x, -4.371\}\right]$$

$$\text{FindRoot}\left[1 + \frac{x^3}{100} == x \, \text{Sin}[x], \{x, -4.370\}\right]$$

$$\text{FindRoot}\left[1 + \frac{x^3}{100} == x \, \text{Sin}[x], \{x, -4.369\}\right]$$

$$\text{FindRoot}\left[1 + \frac{x^3}{100} == x \, \text{Sin}[x], \{x, -4.368\}\right]$$

Out[25]= $\{x \rightarrow -1.10446\}$

Out[26]= $\{x \rightarrow 6.9627\}$

Out[27]= $\{x \rightarrow 8.44329\}$

Out[28]= $\{x \rightarrow -1.10446\}$

8.5 Differential Equations

Let's see how to solve differential equations with *Mathematica*. Suppose we want to find all solutions to the equation

$$y'(x) = 3y(x)$$

In other words, we are looking for a function whose derivative is equal to 3 times itself. The following example shows how we can use **DSolve** to solve this equation.

Example 8.5.1

In[4]:= (* solving a differential equation with DSolve *)
    DSolve[y'[x] == 3 y[x], y[x], x]

Out[4]= $\left\{\left\{y[x] \rightarrow e^{3x} \, C[1]\right\}\right\}$

To use **DSolve** we need to enter three arguments. The first is the differential equation that we want to solve, the second is the function we want to solve for, and finally the third argument is the independent variable. Notice that the answer is given in the form of a replacement rule. In this case, there are infinitely many solutions to the differential equation indicated by the fact that the solution contains the arbitrary constant **C[1]**. If we want a specific solution to the equation, for example, a solution for which $y(2) = \pi$, then this would determine the constant since we would need $\pi = y(2) = Ce^6$, and therefore $C = \pi e^{-6}$.

Instead of having to figure out the value of **C[1]** ourselves, we can have *Mathematica* do it by providing the *initial condition* (for example, $y(2) = \pi$) from the beginning. The next example shows how to do this. Notice that we simply replace the first argument to **DSolve** with a list containing both the differential equation and the initial condition.

Example 8.5.2

```
In[29]:= (* stipulating an initial condition *)
         DSolve[{y'[x] == 3 y[x], y[2] == π}, y[x], x]
```

Out[29]= $\left\{\left\{y[x] \to e^{-6+3\,x}\,\pi\right\}\right\}$

Having the output in the form of a replacement rule can be useful if we want to graph the solution. In Example 8.5.3, we give the name **solution** to the output of **DSolve** and then use the replacement operator to substitute for **y[x]** in the **Plot** function.

On the other hand, there are other operations that would be difficult to accomplish using the replacement rule that we obtain from **DSolve**. For example, suppose that we simply want to verify that the solution is correct. We want to show that $y'(x) = 3y(x)$ and that $y(2) = \pi$. The initial condition can be confirmed without too much trouble as follows. Notice that we first substitute the solution for $y(x)$ and then 2 for x.

Example 8.5.3

```
In[30]:= (* plotting the solution to the differential
         equation *)
         solution = DSolve[
           {y'[x] == 3 y[x], y[2] == π}, y[x], x
         ]
         Plot[y[x] /. solution, {x, 0, 1}]
```

Out[30]= $\left\{\left\{y[x] \to e^{-6+3\,x}\,\pi\right\}\right\}$

Example 8.5.3 (Continued)

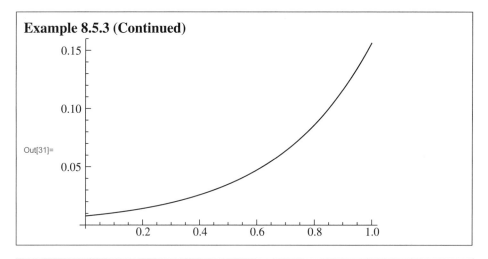

Out[31]=

Example 8.5.4

In[32]:= (* confirming that y[2]=π *)
    y[x] == π /. solution /. x → 2

Out[32]= {True}

We can verify the differential equation in a similar way, as seen below, but it is more awkward. Notice that our first attempt is foiled by the fact that the original output of **DSolve** is delimited by a pair of curly braces. This is because **DSolve** can solve sets of equations that might have multiple solutions and is therefore designed to return a list of lists. We'll see examples of this soon. Our second attempt gets around this problem by taking the first part of the derivative. (We could also have taken the first part of **solution** before we differentiated and used **D[y[x]/.solution[[1]], x]==3 y[x]/.solution**. Try it!)

Example 8.5.5

In[32]:= (* trying to confirm that y'[x]=3 y[x] *)
    D[y[x] /. solution, x] == 3 y[x] /. solution
    D[y[x] /. solution, x][[1]] == 3 y[x] /. solution

Out[32]= $\left\{\left\{3\ e^{-6+3\ x}\ \pi\right\} == 3\ e^{-6+3\ x}\ \pi\right\}$

Out[33]= {True}

But perhaps more awkward is that we cannot simply write **y′[x]/.solution** and instead need to use the **D** function. This is because the replacement rule can only

Mathematica Demystified

be used to replace the expression **y[x]** and therefore will not replace the expression **y′[x]**.

Compare all of this with the following example, where we simply use **y** as opposed to **y[x]** as the second argument to **DSolve**. Doing this causes **DSolve** to return a pure function as the solution and this pure function can be substituted into expressions like **y′[x]**.

Example 8.5.6

```
In[33]:= (* using pure functions *)
        solution = DSolve[{y'[x] == 3 y[x], y[2] == π}, y, x]
        (* verifying the differential equation *)
        y'[x] == 3 y[x] /. solution
        (* verifying the initial condition *)
        y[2] == Pi /. solution
```

$$Out[33]= \left\{ \left\{ y \rightarrow Function\left[\{x\}, e^{-6+3\,x}\,\pi \right] \right\} \right\}$$

Out[34]= {True}

Out[35]= {True}

If we use **DSolve** to produce a pure function as in the last example, we can still graph the solution exactly as we did in Example 8.5.3. Try it out for yourself! Whether it is better to use pure functions or not will depend on what you are trying to do with the solution. But if you understand the difference between pure and nonpure functions you shouldn't have any trouble working with either.

Before leaving the arena of first order equations, let's look at one more example. If we do not specify initial conditions then the general solution will involve an arbitrary constant. In Example 8.5.7, we use **Table** to create a list of particular solutions and then plot all of them at once. Notice that this differential equation produces two different solutions so that the output from **DSolve** is a list with two elements. The solutions are nearly the same: a plus sign is switched to a minus sign to go from the first to the second solution. Also, observe the construction **y[x]/. sol /. C[1]→k** in the **Table** function. We first replace **y[x]** with the solution and then replace **C[1]** with **k** before letting **k** run from −10 to 10 in steps of 1. It might seem that **Table[y[x] /. sol, {C[1], -10, 10, 1}]** would be simpler, but unfortunately this will not work. *Mathematica* complains that "Tag C in C[1] is Protected." We also use **Tooltip** so that as the mouse is moved over different solutions the value of the constant is displayed.

Mathematica can solve higher order differential equations too. The next example solves a fairly simple equation.

Example 8.5.7

```
In[88]:= (* plotting different solutions *)
       (* mouseover displays C[1] *)
       sol = DSolve[y'[x] == (x² - y[x]) / (x + y[x]), y, x]
       (* vary C[1] to produce list of solutions *)
       particularSolutions = Table[
          Tooltip[y[x] /. sol /. C[1] → k, k], {k, -10, 10}
          ];
       (* plot all solutions in list *)
       Plot[particularSolutions, {x, -4, 4}]
```

$$Out[88]= \left\{\left\{y \to \text{Function}\left[\{x\}, -x - \sqrt{x^2 + \frac{2x^3}{3} + C[1]}\right]\right\},\right.$$

$$\left.\left\{y \to \text{Function}\left[\{x\}, -x + \sqrt{x^2 + \frac{2x^3}{3} + C[1]}\right]\right\}\right\}$$

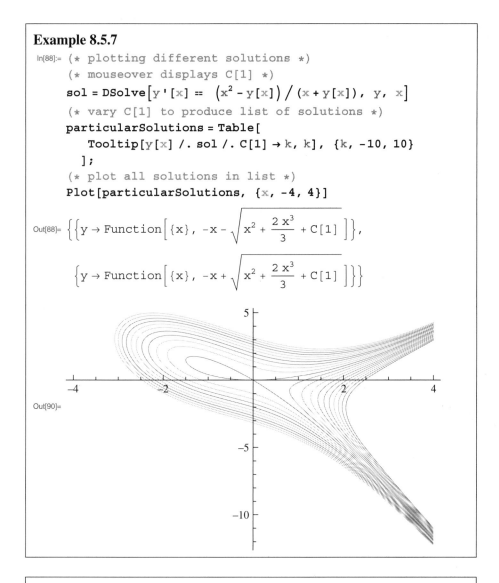

Out[90]=

Example 8.5.8

```
In[450]:= (* solving a second order differential equation *)
        DSolve[y''[x] + y[x] == x, y, x]
Out[450]= {{y → Function[{x}, x + C[1] Cos[x] + C[2] Sin[x]]}}
```

Notice that in this example the solution involved two unknown constants, **C[1]** and **C[2]**. If we were to specify initial values for both y and y' we would get a specific solution. Example 8.5.9 illustrates this.

Mathematica Demystified

Example 8.5.9

```
In[104]:= (* solving a second order differential equation
       with initial conditions for both y and y' *)
     DSolve[
       {y''[x] + y[x] == x, y[0] == 2, y'[0] == 0},
       y, x
     ]
Out[104]= {{y → Function[{x}, x + 2 Cos[x] - Sin[x]]}}
```

We close this section with an example that uses a really cool feature of **Manipulate**. Suppose we want to solve the system of first order differential equations

$$x'(t) = x(t) - 10y(t)$$

$$y'(t) = 15x(t) + y(t)$$

We can do this with **DSolve**. We simply need to list both differential equations as the first argument to **DSolve**. We do this in Example 8.5.10. Notice that the second argument is also a list—a list of the two unknown functions. We are using **x** and **y** rather than **x[t]** and **y[t]** so we will be getting pure functions as answers. The third argument is the independent variable t.

Example 8.5.10

```
In[105]:= (* solving a pair of first order linear
       equations simultaneously *)
     solution = DSolve[
       {x'[t] == x[t] - 10 y[t], y'[t] == 15 x[t] + y[t]},
       {x, y}, t
     ]
```

$$Out[105]= \left\{\left\{x \rightarrow \text{Function}\left[\{t\},\right.\right.\right.$$

$$e^t C[1] \cos\left[5\sqrt{6}\ t\right] - \sqrt{\frac{2}{3}}\ e^t C[2] \sin\left[5\sqrt{6}\ t\right]\right],$$

$$y \rightarrow \text{Function}\left[\{t\},\ e^t C[2] \cos\left[5\sqrt{6}\ t\right] +\right.$$

$$\left.\left.\left.\sqrt{\frac{3}{2}}\ e^t C[1] \sin\left[5\sqrt{6}\ t\right]\right]\right\}\right\}$$

Next, let's plot the solution in the xy-plane. We can use **ParametricPlot** to do this. But, of course, we can only plot a specific solution, and so far we have found the general solution which still contains two arbitrary constants. To plot a specific solution we could first choose specific values for the constants. Before we get started, a good question is, How do the constants relate to the initial values $x(0)$ and $y(0)$? Example 8.5.11 shows us that these initial values of x and y are simply equal to the two constants.

Example 8.5.11

In[66]:= (* finding the initial values in terms of
      the constants C[1] and C[2] *)
    {x[0], y[0]} /. solution

Out[66]= {{C[1], C[2]}}

So, by varying the constants **C[1]** and **C[2]**, we will be varying the initial values $x(0)$ and $y(0)$. Let's plot the solution and use **Manipulate** to vary the initial conditions. But instead of using sliders as controllers, wouldn't it be cool to move the initial point $(x(0), y(0))$ around in the xy-plane by using the mouse? We can do this with the **Locator** feature of **Manipulate**! This is a nice feature of *Mathematica* that allows us to dynamically interact with graphical output! Example 8.5.12 illustrates how to do this.

Let's see how the code works. The body of the **Manipulate** construction consists of just one instruction, **ParametricPlot**, which is then followed by two controllers. The first is **length** and the second is **pt** (which stands for point). The first controller is realized as a familiar slider. But the second controller is of the **Locator** type. This causes a dynamically interactive *locator point* to appear in the graphic which we can then drag with the mouse. **Locator** controls have to be given an initial value, so the controller definition begins with {**pt**,{**x0, y0**}}, where {**x0, y0**} is the initial value of **pt**. In this case, we supply the initial values for the coordinates randomly by using **RandomReal[ ]**. The **RandomReal[ ]** function will return a randomly chosen real number between 0 and 1. So, the overall effect of the controller definition is to define the controller **pt** as a **Locator** with an initial position that has been chosen at random.

If we now look at the **ParametricPlot** function, we can see that the curve to be plotted is given parametrically by $\{x(t), y(t)\}$, where x and y are replaced by the solution to the differential equation and finally the constants **C[1]** and **C[2]** are replaced with the first and second parts of **pt**. The domain of the plot is $0 \le t \le$ length, where the controller **length** is initially set to 0.5, but is allowed to run from 0 to 1. You really need to try this out! The effect of moving the initial point with

Mathematica Demystified

the mouse is quite dramatic as is changing the length of the trajectory by using the slider.

Example 8.5.12

In[106]:= `(* using Locator to Manipulate the initial`
`        conditions while plotting the solution *)`

```
Manipulate[
 ParametricPlot[
  {x[t], y[t]} /. solution /.
   {C[1] → pt[[1]], C[2] → pt[[2]]},
  {t, 0, length},
  PlotRange → {{-2, 2}, {-2, 2}}
 ],
 (* manipulate controllers *)
 {{length, .5, "Length"}, 0, 4},
 {{pt, {RandomReal[], RandomReal[]}}, Locator}
]
```

Out[106]=

A couple of remarks are in order. What happens if you run the length slider all the way down to zero? *Mathematica* will object because the domain of the **ParametricPlot** will then be from 0 to 0. To avoid this it is handy to set the range of **length** to be from 0.001 (or some other small number) to 4. Secondly, you may wish to experiment with the inclusion of the option **PerformanceGoal→"Quality"** in the **ParametricPlot** function. If **length** is set to the maximum (especially if that maximum is increased to 4 or more) and the locator point is moved quickly, the trajectory is seen to be more polygonal. On the other hand, including the option slows down the responsiveness of the locator point.

8.6 Find Out More

As usual, we have barely scraped the surface of what *Mathematica* can do when it comes to solving equations. To find out more, check out the following items and tutorials in the Help Files:

- **Reduce**.
- **Eliminate**.
- tutorial/SolvingEquations.
- tutorial/EquationsInOneVariable which is a section in the larger tutorial/ ManipulatingEquationsAndInequalitiesOverview.
- tutorial/DSolveOverview. This is a *huge* overview. Pay special attention to Working with DSolve—A User's Guide which is a really nice guide.
- tutorial/DifferentialEquations-Basics.

An entire application for tracing out trajectories of differential equations (similar to what we did in Example 8.5.12) exists under the name "Equation Trekker" which is part of a *Mathematica package*. We have generally avoided discussing *Mathematica* packages in this introductory book, but take a look at the tutorial: EquationTrekker/tutorial/EquationTrekker.

Quiz

1. Solve $xy + y - 3 = \frac{2x+y}{3x+4}$ for x in terms of y. Repeat the problem, but solve for y in terms of x.
2. Find the roots of $x^4 + 5x^3 - x + 6$, both exactly and approximately.
3. Plot the roots of $x^6 + x + 1$.

4. Find the value of a where $x^6 + ax + 1$ has repeated roots. What is the least number of distinct roots this polynomial can have?

5. Find all solutions to the following system of linear equations.

$$x - z = 4$$

$$2x + y - 3z = 5$$

6. Find the points where the two curves shown below intersect. The equations of these curves, in polar coordinates, are

$$r = 2\sin(3\theta) \qquad \text{and} \qquad r = 1 + \cos^2(3\theta)$$

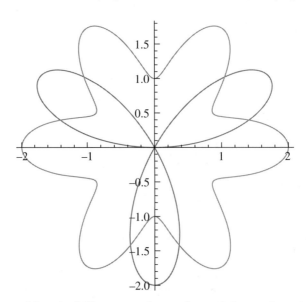

7. Suppose an object is falling near the surface of the earth and that its height off the ground at time t is $x(t)$. If there is no air resistance, and assuming that the acceleration due to gravity is constant at $g = 9.8$ meters per second per second, then x satisfies the following differential equation:

$$x''[t] = -g.$$

Use **DSolve** to find $x(t)$. If the object is dropped from a height of 1000 meter above the ground how long will it take to reach the ground?

8. Let's add in the effect of air resistance to the falling body of the previous problem. Suppose that as the object falls the drag is proportional to the velocity.

This gives an upward force of $kx'(t)$ where k is some negative constant in addition to the downward force of $-mg$ where m is the mass of the object. We now have the differential equation

$$x''(t) = \frac{k}{m}x'(t) - g$$

The constant k/m depends on the size and shape of the falling body. For a skydiver in a spead-eagle position it is about -0.163. If a skydiver jumps out of a plane 3000 meters above the ground, how long will it take her to fall 1000 meters? (She plans to open her chute about 1500 meters off the ground.)

9. Continuing with the skydiver of the last question, what is her *terminal velocity*, that is, the velocity she reaches when the force of drag matches the force of gravity and she no longer accelerates?

10. Change Example 8.5.12 to include three randomly chosen locator points, each providing the initial point of a solution, instead of just one. Can you write the code so it would be trivial to change that to five or ten or twenty points?

CHAPTER 9

Working with Data

One of the really incredible features of *Mathematica* is the access it provides to a number of large data sets. These data sets are stored and regularly updated by Wolfram Research and available over the internet. Simple *Mathematica* functions are provided that can be used to explore and analyze the data. If you take a look at "guide/DataCollections" in the Help Files you'll see that included among the data sets are mathematical data (polyhedra, graphs, knots, and the like), physical and chemical data (elements, chemicals, isotopes, particles, and the like), country and city data, financial data (currency, stocks, and the like), and word data (dictionary, thesaurus, and the like). Using these data sets you can easily render a three-dimensional plot of your favorite molecule, chart the closing share prices of the stocks in your retirement portfolio, or build star charts for the night sky at your location on earth.

In this chapter, we'll explore two of these data sets: CountryData and WordData. When working with large data sets of any kind the necessary tasks are usually the same. We need to import data from some source, select features of the data that interest us, analyze the data, and usually display the data and our analysis in some way (graphical, tabular, and so on), often so that we can present our findings to others.

9.1 Country Data

Using **CountryData** we can access all sorts of information on over 200 countries around the world. Example 9.1.1 shows the basic usage of this function, where we find out the population, area and GDP (gross domestic product) for Sudan, Russia, and Chile. **CountryData** takes two arguments, the first being the name of the country and the second a *property*, where both arguments are in quotes.

Example 9.1.1

```
In[293]:= (* getting information about countries *)
    CountryData["Sudan", "Population"]
    CountryData["Russia", "Area"]
    CountryData["Chile", "GDP"]
```

Out[293]= 4.02185×10^7

Out[294]= 1.70752×10^7

Out[295]= 1.45841×10^{11}

An optional third argument of "Units" can be used to determine what units are being used. In Example 9.1.2, we extract a variety of geographical data about the United States and display the results in a table. The four properties that we are interested in are first given in a list. We then map onto this list a pure function that creates a list of triplets. Each triple contains the property, then the value for the United States, and then the units that are used. Finally, **TableForm** is used to display the output in tabular form.

Example 9.1.2

```
In[298]:= (* tabulating some geographical data for the US *)
    TableForm[
     {#, CountryData["UnitedStates", #],
        CountryData["UnitedStates", #, "Units"]} & /@
      {"Area", "CoastlineLength", "HighestPoint",
        "LowestPoint"}
    ]
```

Out[298]//TableForm=

Area	9.63142×10^6	SquareKilometers
CoastlineLength	19 924.	Kilometers
HighestPoint	MountMcKinley 6194.	Meters
LowestPoint	DeathValley −86.	Meters

To find out what properties are available, enter **CountryData["Properties"]**. To find out what countries are in the data set, enter **CountryData[ ]**. In Example 9.1.3, we find the list of properties, but use **Take** to display only the first twenty. Try using **Length[CountryData["Properties"]]** to find out how many different properties there are.

Example 9.1.3

```
In[304]:= (* the first 20 properties *)
    Take[
     CountryData["Properties"],
     20
    ]
```

```
Out[304]= {AdultPopulation, AgriculturalProducts,
     AgriculturalValueAdded, Airports,
     AlternateNames, AlternateStandardNames,
     AMRadioStations, AnnualBirths, AnnualDeaths,
     AnnualHIVAIDSDeaths, ArableLandArea,
     ArableLandFraction, Area, BirthRateFraction,
     BorderingCountries, BordersLengths, BoundaryLength,
     CallingCode, CapitalCity, CapitalLocation}
```

In Example 9.1.4, we find the ten most wealthy nations (in terms of GDP). We could easily change this to give the ten countries with the most people, land area, and so on, by simply replacing "GDP" with whatever property we want.

Example 9.1.4

```
In[273]:= (* finding the 10 countries with highest GDP *)
    richest = Take[
     Sort[
      {#, CountryData[#, "GDP"]} & /@ CountryData[],
      #1[[2]] > #2[[2]] &
     ],
     10]
```

$$Out[273]= \left\{ \left\{ \text{UnitedStates}, \; 1.31923 \times 10^{13} \right\}, \right.$$
$$\left\{ \text{Japan}, \; 4.43499 \times 10^{12} \right\}, \left\{ \text{Germany}, \; 2.8887 \times 10^{12} \right\},$$
$$\left\{ \text{China}, \; 2.66677 \times 10^{12} \right\}, \left\{ \text{UnitedKingdom}, \; 2.3725 \times 10^{12} \right\},$$
$$\left\{ \text{France}, \; 2.23439 \times 10^{12} \right\},$$
$$\left\{ \text{Italy}, \; 1.848 \times 10^{12} \right\}, \left\{ \text{Canada}, \; 1.27063 \times 10^{12} \right\},$$
$$\left\{ \text{Spain}, \; 1.22501 \times 10^{12} \right\}, \left\{ \text{Brazil}, \; 1.0678 \times 10^{12} \right\} \right\}$$

Let's see how this example works. We begin by using **CountryData[ ]** to return the list of countries. We then map a pure function onto this list to create a list of pairs, where each pair contains the country name and then the GDP. Since we want the ten wealthiest countries, we need to sort this list of pairs. Thus we use **Sort**. But we want to sort on the second element of each pair, so the second argument to **Sort** is the pure function ♯**1[[2]]**> ♯**2[[2]]&**. The countries and their GDPs are now listed in decreasing order of GDP and we use **Take** to take the first 10 elements of this list. If we wanted the 10 poorest countries we could either **Take** the last 10 by using −10, or we could change the ordering function used in **Sort** by replacing the greater than symbol with the less than symbol.

When working with data sets, we often want to display the data in ways that make it easier to see and comprehend. In Example 9.1.5, we display the data we collected in the list **richest** by means of a bar chart. There is a nice function called **BarChart** which is just what we need. **BarChart** is not part of the main *Mathematica* kernal but instead is part of an auxiliary *package*.

In earlier versions of *Mathematica* many functions were grouped together in a variety of different packages. Presumably the idea was that the kernal would contain the most often used core functions and that more infrequently used esoteric functions would be in separate packages that could be loaded in and used as needed. As the development of *Mathematica* proceeds, more and more of the packages are being incorporated into the main kernal. For that reason, and because this book is intended primarily for beginners, we have avoided talking about packages. But, the **BarChart** function is so useful that it would be a shame not to use it!

Example 9.1.5

```
In[306]:= (* plotting the GDP of countries *)
    Needs["BarCharts`"]
    labels =
      Graphics[Text[♯[[1]], Scaled[{1, 1}], {0, 1.54},
          {0, -1}], ImageSize → {20, 80}] & /@ richest;
    BarChart[♯[[2]] & /@ richest,
     BarLabels → labels,
     BarGroupSpacing → .2,
     PlotLabel →
       "GDP of Ten Richest Countries in US Dollars"
     ]
```

Example 9.1.5 (Continued)

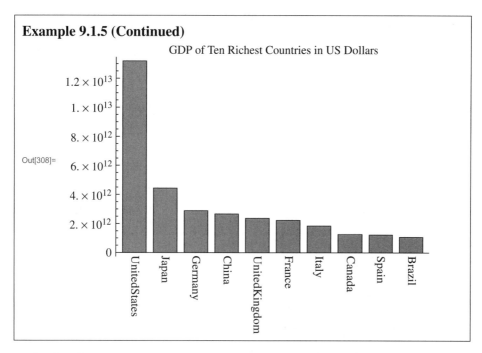

Out[308]=

The first line of Example 9.1.5 is **Needs["BarChart`"]** which tells *Mathematica* to load the BarChart package, thereby making available all the functions in the package. We only need to load the package once. Ignore for a moment **labels** and move onto **BarChart**. This function takes one argument, a list of data, and then several options. The data are the GDP values which we extract from **richest** by mapping the pure function ♯**[[2]]&** onto **richest**. The next three arguments are options to change the appearance of the chart. **BarChart** accepts all the same arguments as **Plot**, for example, plus some extra ones like **BarLabels** and **BarGroupSpacing**. We use **BarLabels** to place the country names under the bars. To do this, we first create a list of the labels which we call **labels**. This is a little tricky as we want the labels to appear vertically and we also want to have them line up under the bars. Basically, we use **Text** to make the labels, but you need to take a good look at the Help Files to see how we have used the three optional arguments of **Text** that follow the country name. The last one for example, {**0,-1**}, is used to rotate the text 90 degrees. Just a word of warning: a lot of futzing around with the second and third arguments to **Text** (coordinates and offset) was needed to make the labels line up nicely with the bars.[1]

---

[1] In fact, **Manipulate** was used to vary the parameter that is now set to 1.54, watching until the alignment of the labels with the bars looked good.

Using **CountryData** we can also draw maps! Cartographic data on each of the countries is available as well as different map projection schemes such as Mercator, Mollweide, and Miller Cylindrical to name just three. In Example 9.1.6, we display the map of Thailand. Here **CountryData["Thailand","Polygon"]** returns a polygon that forms the outline of the country, which we then display using **Graphics**.

Example 9.1.6

```
In[390]:= (* displaying the map of a country *)
Graphics[
 CountryData["Thailand", "Polygon"]
]
```

Out[390]=

We can draw the map of a single country, or a list of countries. In Example 9.1.7, we draw a map of all of South America and further use **Tooltip** to display data about each country as we move the mouse over the map. In this example, the mouse is over Uruguay. Instead of having to enter a list with the names of all the countries of South America, **CountryData["SouthAmerica"]** will return such a list. We then map the pure function **Tooltip[CountryData[♯, "Polygon"], label[♯]]&** onto the list of countries to create the graphics object that we then display with **Graphics**. Try running this example without the first two graphics parameters, **Yellow** and **EdgeForm[Black]**. It will still work but produce a solid black map of South America.

Example 9.1.7

```
In[397]:=  (* using Tooltip to bring up data *)
       Graphics[
        {Yellow,
         EdgeForm[Black],
         Tooltip[
            CountryData[#, "Polygon"],
            label[#]
            ] & /@
          CountryData["SouthAmerica"]
         }
        ]
```

Out[397]=

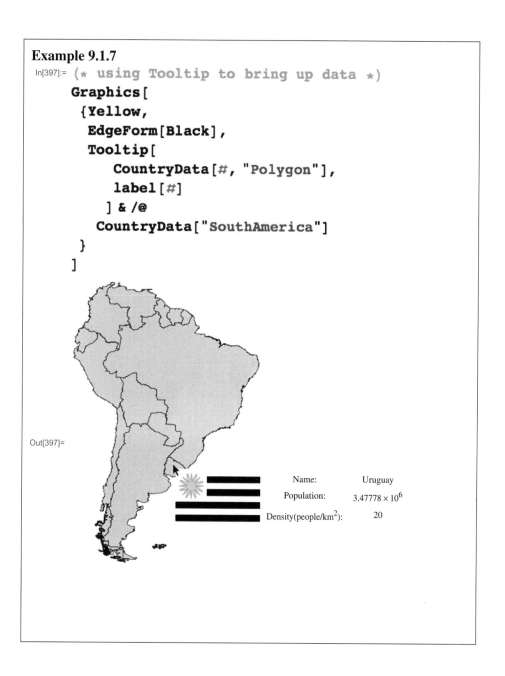

Recall that **Tooltip** takes two arguments. The first is the graphics object and the second is the data that we want to pop up when the cursor is brought over the object. We have used a separate function, called **label** to define the data to be shown for each country. We show this function in Example 9.1.8.

Example 9.1.8

```
In[398]:= (* label for South America Tooltip data *)
    label[country_] := Row[{
      CountryData[country, "Flag"]
      , Text[Grid[
        {{"Name:", country},
         {"Population:",
          CountryData[country, "Population"]},
         {"Density(people/km2):",
          Round[
           CountryData[country, "Population"] /
            CountryData[country, "Area"]
          ]
        }}
      ]]
    }
  ]
```

Here we use **Row** to place the flag and textual data side by side. For the text data on the right we use **Text[Grid[ ]]**, and, of course, the population and population density are obtained by using **CountryData**. We divide the population by the area to get the number of people per square kilometer and round this quotient off to the nearest integer using **Round**.

We close this section by showing how we obtained the population data used in Example 2.8.2. We begin by getting population data over a range of years as shown in Example 9.1.9. Notice that the first element of each data point is a time in the format {*year, month, day, hour, minute, second*}. In Example 2.8.2, we only want data at 5-year intervals. It would be nice if **CountryData** took an optional stepsize in the year range, but it doesn't. So, we'll use **Select** to pick out years that are divisible by 5. That will almost create the list we want. But, we only need the year from each time list, and we want the population in millions. So we map the pure function {#[[1, 1]], #[[2]]/10^6} **&** onto the list.

Example 9.1.9

In[82]:= (* getting data over a range of time *)
        CountryData[
         "UnitedStates",
         {"Population", {2000, 2005}}
        ]

Out[82]= $\{\{\{2000, 1, 1, 0, 0, 0\}, 2.84857 \times 10^8\},$
        $\{\{2001, 1, 1, 0, 0, 0\}, 2.87837 \times 10^8\},$
        $\{\{2002, 1, 1, 0, 0, 0\}, 2.90832 \times 10^8\},$
        $\{\{2003, 1, 1, 0, 0, 0\}, 2.93837 \times 10^8\},$
        $\{\{2004, 1, 1, 0, 0, 0\}, 2.96844 \times 10^8\},$
        $\{\{2005, 1, 1, 0, 0, 0\}, 2.99846 \times 10^8\}\}$

Example 9.1.10

In[85]:= (* selecting every fifth year and formating
        the data *)
        usData = {#[[1, 1]], #[[2]] / 10^6} & /@
         Select[
          CountryData[
           "UnitedStates",
           {"Population", {1970, 2005}}
          ],
          Mod[#[[1, 1]], 5] == 0 &
         ]

Out[85]= {{1970, 210.111}, {1975, 220.165},
        {1980, 230.917}, {1985, 243.063},
        {1990, 256.098}, {1995, 270.245},
        {2000, 284.857}, {2005, 299.846}}

We have barely scratched the surface of what can be done with **CountryData**. You'll find lots of interesting examples in the Help Files that can lead you deeper into the data set.

9.2 Word Play

One of the large data sets available through *Mathematica* is a dictionary of English words. In this section, we'll explore several functions that are provided to make use of the dictionary.

In Example 9.2.1, we see how to use the function **DictionaryLookup** to find all words in the dictionary that begin with the letter "a" and end with the letter "k."

Example 9.2.1

```
In[11]:= (* find all words that start with a and
         end with k *)
         a2kWords = DictionaryLookup["a" ~~ ___ ~~ "k"]

Out[11]= {aardvark, aback, aftershock, airlock, airsick,
         alack, almanack, amok, anorak, antiknock,
         antitank, apparatchik, applejack, ark,
         artwork, ask, asterisk, attack, auk, awestruck}
```

Here we have provided a *pattern* as the single argument to **DictionaryLookup**. The function then finds all words in the dictionary that match the pattern. The three underscores will match any sequence of zero or more *Mathematica* expressions. The double tildes are used to join together successive parts of the pattern.[2] Thus our pattern will match any word that starts with "a," ends with "k," and has any number of letters (including none) in between. Since "ak" is not a word, all the matches have three or more letters.

If we want to allow only one letter between "a" and "k," we can use the single underscore instead of the triple underscore. A single underscore will match only a single letter. In Example 9.2.2, we illustrate two ways to find all three-letter words that start with "a" and end with "k." The first is to use the single underscore pattern and the second is to use the **Select** function to extract the three-letter words from the list of all words that start with "a" and end with "k." In Example 9.2.1, we named this list **a2kWords**. Now we **Select** from this list using the pure function **StringLength[♯]==3 &**.

We can even use a pattern which specifies letters to *not* match. Example 9.2.3 uses **Except** to find all words that contain a "q" that is *not* followed by a "u."

---

[2]The double tilde $\sim\sim$ is similar, but not quite the same, as $<>$, which is used to concatenate strings. If str_1 and str_2 are two strings, then $str_1\sim\sim str_2$ and $str_1<>str_2$ will both concatenate the two strings. But $\sim\sim$ can be used in greater generality.

Example 9.2.2

In[23]:= (* two ways to find all three-letter words
        that start with a and end with k *)
     DictionaryLookup["a" ~~ _ ~~ "k"]
     Select[a2kWords, StringLength[#] == 3 &]

Out[23]= {ark, ask, auk}

Out[24]= {ark, ask, auk}

Example 9.2.3

In[5]:= (* find all words that have a q NOT followed
       by a u *)
     DictionaryLookup[___ ~~ "q" ~~ Except["u"] ~~ ___]

Out[5]= {Chongqing, Iqaluit, Iqbal,
      Iraqi, Iraqis, Qiqihar, qwerty, Urumqi}

In addition to the single and triple underscore, there is also a double underscore that can be used to stand for any sequence of one or more *Mathematica* expressions. Thus **DictionaryLookup["a"~~_ _ ~~"k"]** and **DictionaryLookup ["a"~~_ _ _ ~~"k"]** will yield the same results since "ak" is not a word. But using "o" and "f" instead of "a" and "k" will give different results.

Finding all words that are a certain length and have certain letters in certain positions is just what we need for the Sunday Crossword Puzzle! In Example 9.2.4, we give two examples of finding all words that might be needed for "21 Across" or "52 Down."

A popular word puzzle that appears in many newspapers requires the player to unscramble a word. For example, we might need to rearrange the letter of "toffes" to form a real word. We can easily write a program that will do this! First we'll use **Permutation** to rearrange the letters in all possible ways, and then use **DictionaryLookup** to see which of these rearrangements are real words. Example 9.2.5 illustrates the **Characters** and **Permutation** functions. The first will convert a string into a list of its characters, and the second will permute the elements of any list in all possible ways.

In Example 9.2.6, we continue with the previous example, by first mapping **StringJoin** over the list of permuted letters in order to reform them into words, and then mapping **DictionaryLookup** over the list of words to see which are in the dictionary. (Note that the patterns in this case are exact words, with no underscores of any kind to allow for multiple matches.)

Example 9.2.4

```
(* all five-letter words that start with "f"
 and end with "nd" *)
DictionaryLookup["f" ~~ _ ~~ _ ~~ "nd"]
```

{fiend, found, frond}

```
(* all words that start "ab" and have five
 letters *)
Select[DictionaryLookup["ab" ~~ ___],
 StringLength[#] == 5 &]
```

{abaci, aback, abaft, abase, abash, abate,
 abbes, abbés, abbey, abbot, abeam, abets,
 abhor, abide, abler, ables, abode, abort,
 about, above, abuse, abuts, abuzz, abyss}

Example 9.2.5

```
In[29]:= (* splitting a word into its letters *)
    Characters["par"]
```

Out[29]= {p, a, r}

```
In[30]:= (* permuting the letters in all possible
    ways *)
    Permutations[Characters["par"]]
```

Out[30]= {{p, a, r}, {p, r, a}, {a, p, r},
    {a, r, p}, {r, p, a}, {r, a, p}}

Example 9.2.6

```
In[42]:= (* split a word into its characters,
    permute characters in all ways,
    rejoin the letters to a string,
    then look it up in the dictionary *)
    DictionaryLookup /@
     StringJoin /@
     Permutations[Characters["star"]]
```

Out[42]= {{star}, {}, {}, {}, {}, {}, {},
    {}, {}, {tars}, {}, {}, {}, {}, {}, {},
    {}, {arts}, {}, {}, {}, {}, {}, {rats}}

A nice thing to do to Example 9.2.6 would be to use **Select** to save only the words that occur, throwing out the empty lists. Another nice variation is to use

Permutations[*list, n*] which will give all permutations of at most *n* letters. This will allow you to find all words that can be made from the given letters, not just words of the same length.

There is a great deal of information available about the word data set and we can access this information with the **WordData** function. Example 9.2.7 shows a few examples. Using **WordData** we can find definitions, synonyms, antonyms, and more for any given word. We can also find phrases that use the word, as seen in the last cell of Example 9.2.7. To find all the properties that are available in

---

Example 9.2.7

In[178]:= (* using WordData to investigate a word *)
    WordData["vault"]

Out[178]= {{vault, Noun, Jump}, {vault, Noun, Roof},
    {vault, Noun, Sepulcher}, {vault, Noun, Strongroom},
    {vault, Verb, Bound}, {vault, Verb, Overleap}}

In[151]:= **WordData["vault", "PartsOfSpeech"]**

Out[151]= {Noun, Verb}

In[152]:= **WordData["vault", "Definitions"]**

Out[152]= {{vault, Noun, Jump} →
       the act of jumping over an obstacle,
     {vault, Noun, Roof} →
      an arched brick or stone ceiling or roof,
     {vault, Noun, Sepulcher} →
      a burial chamber (usually underground),
     {vault, Noun, Strongroom} →
      a strongroom or compartment (often made
        of steel) for safekeeping of valuables,
     {vault, Verb, Bound} → bound vigorously,
     {vault, Verb, Overleap} →
      jump across or leap over (an obstacle)}

In[158]:= **WordData["vault", "Synonyms", "List"]**

Out[158]= {bank vault, burial vault, hurdle, overleap}

In[163]:= **WordData[___ ~~ "vault" ~~ ___, "Lookup"]**

Out[163]= {bank vault, barrel vault, burial vault, fan vaulting,
      groined vault, pole vault, pole vaulter,
      pole vaulting, ribbed vault, vault, vaulted,
      vaulter, vaulting, vaulting horse, vault of heaven}

WordData, enter **WordData[All, "Properties"]**. The Help Files contain a wealth of information on **WordData** and we encourage you to peruse the examples there.

9.3 Graphs

When working with large data sets, we often want to uncover connections among the data and then somehow display these relationships. Sometimes *graphs* are just what we need. Recall that a graph is a set of points called *vertices* which are connected by line segments called *edges*. Typically the vertices will correspond to data points and the edges to connections between data. In this section, we'll look at a simple, but quite interesting, example using the English dictionary mentioned in the last section. Before we do that, let's see how to display a graph.

In Example 9.3.1, we have used **GraphPlot** to display a graph that has six vertices and seven edges. The main argument to **GraphPlot** is the list $\{1 \to 2, 2 \to 3, 2 \to 4, 2 \to 5, 5 \to 3, 5 \to 4, 4 \to 6\}$ which indicates that vertex 1 is connected to vertex 2, that 2 is connected to 3, and so on. We have also used two options with **GraphPlot**: **VertexLabeling** and **DirectedEdges**. Without the first of these, the vertices would appear simply as unlabeled points and without the second the edges would be drawn without arrowheads. Try this example without the options to see how it looks.

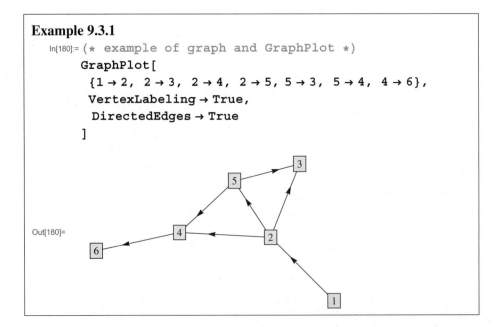

Example 9.3.1

```
In[180]:= (* example of graph and GraphPlot *)
        GraphPlot[
          {1 → 2, 2 → 3, 2 → 4, 2 → 5, 5 → 3, 5 → 4, 4 → 6},
          VertexLabeling → True,
          DirectedEdges → True
        ]

Out[180]=
```

In 1877, Lewis Carroll[3] invented a word game that he originally called Word-Link. The idea is to find a chain of words that connects two given words, with each word in the chain differing from the words before and after it by a single letter. For example, we can connect "ape" to "man" with the chain: ape, apt, opt, oat, mat, man.

Notice that at each step, we change a single letter to go from one word to the next. It's fun to pick two words and try to find a chain between them. Can you find a chain from "good" to "evil"? Is the chain from "ape" to "man" shown above the shortest chain?

We can form an interesting graph by using the set of all words as vertices and then connecting two words with an edge if they differ by a single letter change. Thus the chain from "ape" to "man" becomes an *edge path*, or more simply a *path*, inside this graph. If we construct the graph, there are well-known algorithms that we can then use to find an edge path from any one vertex to any other (provided one exists of course). So we'll be able to write a program that can find word-links between any two given words!

Let's focus on three-letter words and form the associated word-link graph. This is going to be a pretty big graph because there are lots of three-letter words. In the next example we use **DictionaryLookup** to find the set of all three-letter words which we then name **wordSet**.

Example 9.3.2

```
In[186]:= (*  get all words with three letters *)
    wordSet = Select[
        DictionaryLookup[],
        StringLength[#] == 3 &
        ];
    Length[wordSet]

Out[187]= 923
```

Using **DictionaryLookup** without any arguments will return all the words in the dictionary. We then use **Select**, together with the **StringLength** function to select those words that have length three. Notice that there are 923 three-letter words in the dictionary. An alternative way to have defined **wordSet** would have been **wordSet= DictionaryLookup[_~~_~~_];**. This is perhaps simpler, but harder to change if we want six-letter words instead.

---

[3]Lewis Carroll (1832–1898) was the pen name of Charles Dodgson. Best known for writing *Alice's Adventures in Wonderland* and *Through the Looking-Glass*, Dodgson was an accomplished mathematician and logician.

The next thing we need to do is find all the edges between the words. For this we'll need a test to see if two words are related by a single letter change. Example 9.3.3 shows the module **edgeQ**[4] that will decide if two words differ by a single letter. We first see if the words have the same length. If not, the function returns False. If they do have the same length then we begin by splitting each word into a list of its characters. The characters are then checked one by one to see if they agree or disagree. When we are done checking, the function returns true if and only if the number of letter disagreements is exactly one. (Our test is a little inefficient since it keeps checking letters even after two disagreements have been found. For long words this could be a real waste of time, but for three- or four-letter words it's not a big deal.)

Example 9.3.3

```
In[193]:= (* gives True if the two words differ by one letter,
    False otherwise *)
edgeQ[word1_, word2_] := Module[
  {list1, list2, disagree = 0},
  If[StringLength[word1] ≠ StringLength[word2],
   False,
   list1 = Characters[word1];
   list2 = Characters[word2];
   Do[
    If[list1[[i]] ≠ list2[[i]], disagree++],
    {i, 1, Length[list1]}]
   ];
   If[disagree == 1, True, False]
  ]
```

Now that we can decide if two words should be joined by an edge, we are ready to build the graph. We'll consider every possible pair of words, see if they should be connected by an edge, and if so, record the edge. Example 9.3.4 shows how to do this. We could use the list **wordSet** as our list of vertices, but it is convenient to use the list of consecutive integers 1, 2, 3, ... as the vertices. The first line defines the vertex list by using the **Range** function to create a list of consecutive integers starting from 1 and containing as many integers as the number of words. Next we initialize the set **edges** to be the empty list. As we check each pair of words, if we

---

[4] We use a name that ends in Q because *Mathematica* functions which test a True/False question tend to end in Q. Try entering ?*Q to see what functions end in Q.

find a pair that should have an edge between them, we will add that pair to the list of edges. For example, "bad" and "bar" should be connected by an edge. These are words 60 and 65, respectively, in the word list, so we add the pair {60, 65} to the set of edges. Notice how the nested **Do** loops will run through all possible pairs of words from **wordSet**. When we are done there are 5317 edges in the graph. Try evaluating **Take[edges, 100]** to look at the first 100 edges in the graph.

Example 9.3.4

```
In[8]:= (* build the Word-Link graph *)
    vertices = Range[ Length[wordSet]];
    edges = {}; (* initialize edge set to be empty *)
    Do[
     Do[
       (* test each pair of words *)
      If[
        edgeQ[wordSet[[i]], wordSet[[j]]],
        AppendTo[edges, {i, j}]
       ]
       , {j, i + 1, Length[wordSet]}
      ],
      {i, 1, Length[wordSet] - 1}
     ]
    Length[edges]
Out[10]= 5317
```

Let's use **GraphPlot** to view the graph! Since our edges are presently pairs of vertex numbers, we will need to convert this information to the list of vertex numbers connected by arrows as in Example 9.3.1. We do this by mapping the pure function ♯[[1]] → ♯[[2]] **&** onto the list **edges** as seen in Example 9.3.5.

Whoa! What a mess! But even though there are 5317 edges connecting 923 vertices we can still see some important qualitative features about these words by looking at the graph. The first thing we see is that the graph is not *connected*. There are pairs of vertices that are not connected by any edge path. If we were to start with such a pair of words we would not be able to connect them by a Word-Link chain. However, most of the words are in a single connected piece of the graph, with only a few words lying outside the main piece.

It would be nice to show the word corresponding to each vertex. We can do this if we map the pure function **wordSet[[♯[[1]]]]** → **wordSet[[♯[[2]]]]** **&** onto the set of edges (and also use the option **VertexLabeling** → **True**). This will create the list of arrow-pairs using the actual word values rather than just the vertex numbers. You should try it, but the results are rather disappointing as the labels all pile up on top of one another. Unfortunately, we can't just visually inspect this graph to find Word-Links between different words. It's just too big and complicated.

To find a shortest path between two words in the Word-Link graph, we need to remove edges until we obtain a *tree* which is *rooted* at one of the words and which furthermore connects each word to the root word with a minimal length edge path. A *tree* is a graph that contains no loops, that is, edge paths that begin and end at the same vertex without traversing any edge twice. Our graph definitely contains loops. For example, the sequence "ape," "apt," "act," "ace," "ape" forms a loop of four edges that starts and ends at "ape." In this loop, there is not a unique path from "act" to "ape"—we can go either via "ace" or via "apt." But in a tree, it is not hard to see that there is always a unique edge path from any one vertex to any other (assuming, of course, that they lie in the same connected piece of the tree). We can always remove edges from any graph until only a tree remains. We simply need to break every loop. If we consider only the four words "ape," "apt," "act," and "ace," we can break the loop by removing any one of the four edges.

Example 9.3.5

In[222]:= (* all three-letter words with edges connecting
        words that differ by a single letter. We
        need to convert our list of edges to the
        list of arrow-pairs to use Graph Plot *)
GraphPlot[#[[1]] → #[[2]] & /@ edges]

Out[222]=

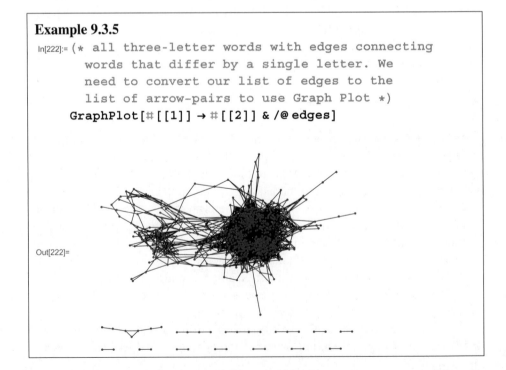

What we want to do is first pick some word and designate it as the root. We'll then only include enough edges of the Word-Link graph to form a tree. Finally, we want each edge path in this tree from any word x to the root to be a minimal length path. There should not be a shorter path between x and the root in the entire Word-Link graph. For example, in Example 9.3.1, if 1 is the root, we should remove the edges from 5 to 4 and from 5 to 3. What remains is a tree and the edge paths from each vertex to vertex 1 that lie in the tree are as short as possible. Notice that we do not want to keep the edge from 5 to 3 and eliminate the edge from 2 to 3. This would still be a tree, but the path from 1 to 3 in the tree would now have length 3, which is not as short as the shortest path from 1 to 3 in the original graph.

We provide a module called **spanningTree** in Example 9.3.6 that will find the desired tree once the root vertex is specified. Unfortunately, this is a pretty complicated function! In fact, it is easily the most complicated function in this book. Rather than explaining how it works,[5] let's just take it for granted and see what we can do with it.

In Example 9.3.7, we first use the **Position** function to find the vertex number corresponding to the word "ape." We see that it is the 36th word in the list **wordSet** and so is vertex number 36. Next we let **tree** be the output of **spanningTree**. The function takes three arguments: the list of vertices, the list of edges, and finally the number of the vertex that is to serve as the root. Finally, we display the first 20 elements of **tree**. In general, **spanningTree** returns a list of triples. Each triple represents an edge and is of the form $\{v, d, w\}$ where v is a vertex connected to vertex w, and d is the distance of vertex v from the root, that is, the number of edges in the edge path from v to the root. Recall that 36 is the root in this case. The first 10 edges in **tree** connect vertices to the root, and so $d = 1$ for each of them. However, the 11th edge in **tree** connects vertex 7 to vertex 6. Since vertex 6 is connected to the root, vertex 7 is now a distance of two from the root.

---

[5]Here is a brief description for the ambitious reader. We first build a list called **neighbors** which in the ith position contains a list of all vertices adjacent (connected by an edge to) vertex number i. We then prepare to enter a **While** loop by initializing two sets of vertices: **leaves** and **freeVertices**. Initially, **leaves** consists of the **neighbors** of **root**, and **free vertices** consists of all vertices except **root** and **leaves**. We also initialize the **tree**, which starts out as the set of all edges that connect the **leaves** to the **root**. Each element of **tree** is a triple $\{v, d, w\}$, where vertex v is connected to vertex w and v is a distance d from the root. In the **While** loop we repeatedly do the following: we go through each vertex v in **freeVertices**, find all its neighbors which are also in the set of **leaves**, and, if there are any, we let w be the first such vertex. We then add the edge connecting v to w to the **tree**. The distance to v is one more than the distance to w. With each pass through the loop, the set of newly formed leaves must replace the old set of leaves, and the set of **freeVertices** must be made smaller. The process continues until no new leaves are found. In the end, **freeVertices** will be empty if and only if the original graph is connected.

Example 9.3.6

```
In[12]:= spanningTree[vertices_, edges_, root_] := Module[
        {neighbors, x, y, freeVertices, leaves, tree,
         level = 1, continue = True, newLeaves, v, W, w},
        neighbors = Table[{}, {Length[vertices]}];
        Do[
         {x, y} = edges[[i]];
         AppendTo[neighbors[[x]], y];
         AppendTo[neighbors[[y]], x],
         {i, 1, Length[edges]}
        ];
        neighbors = Sort /@ neighbors;
        freeVertices = Complement[vertices, {root}];
        leaves = neighbors[[root]];
        freeVertices = Complement[freeVertices, leaves];
        tree = Table[{leaves[[i]], 1, root},
          {i, 1, Length[leaves]}];
        While[continue,
         newLeaves = {};
         Do[
          v = freeVertices[[i]];
          W = Intersection[neighbors[[v]], leaves];
          If[W ≠ {},
           w = W[[1]];
           AppendTo[newLeaves, v];
           AppendTo[tree, {v, level + 1, w}]
          ]
          , {i, 1, Length[freeVertices]}
         ];
         If[newLeaves ≠ {},
          freeVertices = Complement[freeVertices,
            newLeaves];
          leaves = newLeaves;
          level++;
          continue = True,
          continue = False
         ]
        ];
        tree
       ]
```

Example 9.3.7

In[13]:= (* find the vertex number corresponding to ape *)
```
Position[wordSet, "ape"]
```

Out[13]= {{36}}

In[15]:= (* use spanningTree to build tree rooted at ape *)
```
tree = spanningTree[vertices, edges, 36];
```

In[18]:= (* the first 20 elments of tree *)
```
Take[tree, 20]
```

Out[18]= {{6, 1, 36}, {14, 1, 36}, {23, 1, 36}, {37, 1, 36},
    {39, 1, 36}, {48, 1, 36}, {53, 1, 36}, {56, 1, 36},
    {57, 1, 36}, {600, 1, 36}, {7, 2, 6}, {13, 2, 37},
    {15, 2, 14}, {22, 2, 23}, {25, 2, 23}, {26, 2, 23},
    {33, 2, 37}, {38, 2, 39}, {40, 2, 39}, {41, 2, 39}}

Admittedly, the **spanningTree** function is a bit complicated. But now that we have built the tree, let's use **GraphPlot** to plot it. It is still pretty big, so let's only plot those words that are at most a distance of three from the root. Example 9.3.8 shows how to do this. Since the second entry in each element of **tree** is the distance to the root, we first use **Select** with the pure function ♯[[2]]≤3 & to extract the portion of the tree with vertices 1, 2, or 3 edges away from the root. We then plot the tree just as in Example 9.3.5 except that we use the pure function **wordSet[[♯[[1]]]]** → **wordSet[[♯[[2]]]]** & rather than the pure function ♯[[1]] → ♯ [[2]] & so that the words themselves will label the vertices.

In Example 9.3.8, you should be able to make out the path "ape," "apt," "opt," "oat," which we can continue to "man" in two more steps. Still, we don't need to find paths by visually inspecting the tree! It is now a simple matter to write a function that will find the path from any word in the tree to the root. We do this in Example 9.3.9. The function begins by using **Select** to find the **edge** in the **tree** that starts at the given **word**. We then find the **edge** that starts where this **edge** ends and

---

Example 9.3.8

In[19]:= (* plot all vertices that are a distance of 3
    or less from the root *)
```
GraphPlot[
  wordSet[[♯[[1]]]] → wordSet[[♯[[3]]]] & /@
    Select[tree, ♯[[2]] ≤ 3 &],
  VertexLabeling → True
  ]
```

Example 9.3.8 (Continued)

Out[19]=

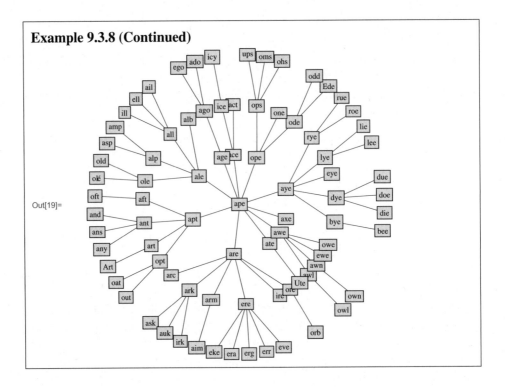

Example 9.3.9

```
In[35]:= (* find path in the tree from word to root *)
    path2root[tree_, word_] := Module[{path = {}, edge},
      edge =
      Select[tree, wordSet[[#[[1]]]] == word &][[1]];
      While[
       edge[[2]] > 1,
       AppendTo[path, wordSet[[edge[[1]]]]];
       edge = Select[tree, #[[1]] == edge[[3]] &][[1]];
      ];
      Join[path,
       {wordSet[[edge[[1]]]], wordSet[[edge[[3]]]]}
      ]
     ]

In[36]:= (* find the path from man to ape *)
    path = path2root[tree, "man"]

Out[36]= {man, mat, oat, opt, apt, ape}
```

CHAPTER 9   Working with Data 261

so on until we reach the **edge** that ends at the **root**. (We use the distance element of each **edge** to know when to stop.) As we find the edges we append them to **path**. When the **While** loop ends we still need to **Join** the last two words to the **path**. You should know that no safeguarding against entering a word that is not in the tree takes place! If you enter a word that is not in **wordSet**, or a word that is in **wordSet** but not in the tree (because the original graph was not connected), the program will fail.

9.4 A Glimpse of Other Data Sets

Mathematica includes far too many large data sets for us to delve into each one. We close this section with just a hint of what else is available.

Suppose we want to follow the progress of some stocks that we own. In Example 9.4.1, we use **DateListPlot** to plot the closing share price of Apple Computer since January 1, 2002. Don't you wish you had bought Apple stock 6 years ago!

Example 9.4.1

```
In[55]:= (* plot daily closing price of Apple
    Computer stock *)
  DateListPlot[
    FinancialData["AAPL", "January 1 2002"],
    Joined → True
  ]
```

Out[55]=

Using **FinancialData** you can find out all sorts of information for publicly traded companies such as share prices, dividends, market capitalization, price-to-earning ratios, and so on. We'll let you read about **DateListPlot** and **FinancialData** in the Help Files.

The data sets, **ChemicalData** and **ElementData**, contain a wealth of physical data. In Example 9.4.2, we find the elements contained in the pain reliever ibuprofen and we also plot the molecular structure. As with other three-dimensional plots, we can grab the graphic with the mouse and rotate it. In Example 9.4.3, we find out a few facts about lead and gold. Again, you should check out the Help Files to see what else is contained in these two data sets and how you can access the data.

Example 9.4.2

In[3]:= (* using ChemicalData to find out about Ibuprofen *)
    ChemicalData["Ibuprofen", "ElementTypes"]

Out[3]= {H, C, O}

In[9]:= ChemicalData["Ibuprofen", "MoleculePlot"]

Out[9]=

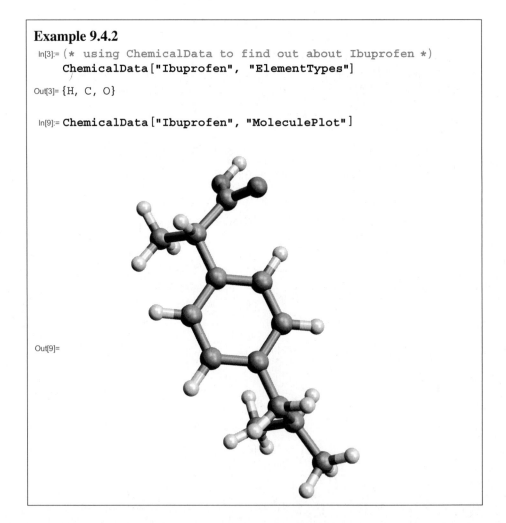

Example 9.4.3

In[12]:= (* using ElementData to find physical
      properties of elements *)
      ElementData["Lead", "MeltingPoint"]

Out[12]= 327.46

In[11]:= **ElementData["Gold", "AtomicWeight"]**

Out[11]= 196.96655

As a final glimpse of what's available, we find the fifteen brightest stars as seen from the earth and then look up their coordinates (right ascension and declination). Since the position of the sun is not fixed with respect to the distant stars, it does not have a constant right ascension and declination. Hence the data is reported as "Missing." Using these coordinates you could, for example, build your own star charts for the night sky in your location.

Example 9.4.4

In[19]:= (* listing the 15 brightest stars *)
      top15 = Take[AstronomicalData["Star"], 15]

Out[19]= {Sun, Sirius, Canopus, Arcturus,
      RigelKentaurusA, Vega, Capella,
      Rigel, Procyon, Achernar, Betelgeuse,
      Hadar, Altair, Acrux, Aldebaran}

In[20]:= (* obtaining the coordinates of the
      15 brightest stars *)
      {AstronomicalData[#, "RightAscension"],
      ** AstronomicalData[#, "Declination"]} & /@**
      top15

Out[20]= {{Missing[Variable], Missing[Variable]},
      {6.75257, -16.7131}, {6.39919, -52.6957},
      {14.2612, 19.1873}, {14.6614, -60.8351},
      {18.6156, 38.783}, {5.27814, 45.999},
      {5.2423, -8.20164}, {7.65515, 5.22751},
      {1.62854, -57.2367}, {5.91952, 7.40704},
      {14.0637, -60.373}, {19.8463, 8.86738},
      {12.4433, -63.0991}, {4.59867, 16.5098}}

9.5 Find Out More

Once again, we have covered only the tip of the iceberg. The following guides, tutorials, and reference pages in the Help Files are definitely worth a look.

- guide/DataCollections—describes the different data sets stored by Wolfram Research.
- ref/CountryData—*lots* of info on how to use this data set.
- guide/DataVisualization—useful introduction to all the different ways to display data.
- BarCharts/guide/BarChartsPackage—you need this to learn how to use **BarChart**.
- PieCharts/guide/PieChartsPackage—you need this to learn how to use **PieChart**.
- `http://en.wikipedia.org/wiki/WordLadder`—a nice place to learn more about Lewis Carroll's game Word-Link.

Quiz

1. What are the 10 most densely and 10 least densely populated countries?
2. Make a scatter plot of infant mortality rate versus GDP for the countries in the **CountryData** data set.
3. Find the population of the world by adding together the population of all the countries in the **CountryData** data set. Make a pie chart that displays population of the 10 largest countries. The pie should have 11 slices: one for each of the 10 biggest countries and one for "other countries" (all the others taken together).
4. Make a map of Africa and use **Tooltip** so that as the mouse is moved over each country on the map information about that country pops up.
5. How many words are in the *Mathematica* dictionary?
6. Find all words that start with "b" and end with "w."
7. The letters of "post" can be rearranged to form six different words. (You can use Example 9.2.6 to verify this!) Are there any four-letter words whose letters can be rearranged to form more than six different words?
8. Rework Examples 9.3.2 to 9.3.9 for four-letter words. Find a shortest Word-Link from "lead" to "gold."

9. Find a shortest Word-Link from "good" to "evil." (Be careful!)

10. The Collatz function $g[n]$ that we discussed in Chap. 5 gives rise to a graph with directed edges, where each integer n is connected by an edge to $g[n]$. Write a program to draw the graph, including all vertices up to a given value. For example, the graph for all vertices up to eight is shown below.

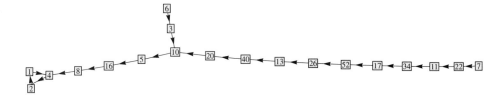

CHAPTER 10

Dynamical Systems and Fractals

In this chapter we are going to learn how to draw beautiful pictures of fractals, strange sets with some pretty amazing properties. Fractals arise naturally in the study of *dynamical systems* so we'll begin our discussion there. A dynamical system is simply a system that is changing as opposed to a *static* system. So everything from the solar system to the flow of blood in our arteries to the pendulum of a clock qualifies as a dynamical system. In order to approach the subject mathematically we usually focus on a single function, or a family of functions, and ask what happens if we repeatedly apply the same function over and over. This simple setup will lead us to *Julia Sets* as well as the famous *Mandelbrot Set*. Once we see how to draw these fractals we'll also learn how to make movies with *Mathematica*.

10.1 Iterating a Function

Suppose we start with a function $f(x)$ and create a "feedback loop" by taking the output of the function and feeding it back in as input. What will happen? For example, suppose $f(x) = x^2$. If we start with an input of 2 we find that the output of f is 4 since f squares the input. If we then square 4 we get 16, squaring 16 gives 256, and so on. Starting from the *initial value*, or initial input of 2, yields the sequence

$$2, 4, 16, 256, 65536, \ldots$$

We can think of this as an *infinite sequence* that goes on without end because there is no limit to how many times we can keep repeating, or *iterating*, the function. Starting with any number x_0 the sequence

$$x_0, f(x_0), f(f(x_0)), f(f(f(x_0))), \ldots$$

obtained by continuing to apply the function f is called the *orbit of x_0 under f*.

The word *orbit* is used to describe the path of a planet as it travels around the sun, and it makes sense to use the word here for the following reason. Imagine that we are studying some big complicated system of objects, perhaps the sun and all the objects that travel around it: the planets, asteroids, comets, and so on. Let's assume that if we knew the mass, positions, and velocities of all the objects at a given moment in time, then we could calculate (at least in theory!) where they will all be located at the next moment in time. That calculation would presumably be incredibly complicated and probably take a really long time to carry out, but imagine that we could do it. Let's name the function that carries out the calculation f. So we would measure the *state* of the solar system at some time (the position of all the objects and their velocities, etc.), enter that state into the function f, and get out the next state of the system. If we repeatedly apply f over and over to some initial state, then the successive states as determined by f would track the future of the solar system, including the orbits of all objects around the sun. So if we start with any function f and follow the successive states, or values, of any initial value, we'll call that the *orbit of the initial value*.

Thinking about the solar system suggests a more spatial or geometric viewpoint rather than a strictly numerical viewpoint and we will be carrying this perspective throughout the chapter. In particular, it makes sense to refer to numbers as points because of the one-to-one correspondence between real numbers and points on the (number) line. So instead of talking about the orbit of a number, we could, and often will, talk about the orbit of a point.

Mathematica has a built-in function called **NestList** that computes the orbit of an initial number (or point) under any function f. Our first example shows the first 6 elements in the orbit of 2 under the squaring function.

Example 10.1.1

```
(* following an orbit of a function *)
f[x_] := x^2
NestList[f, 2, 5]
```

Out[15]= {2, 4, 16, 256, 65536, 4294967296}

The first thing we have done in this cell is define the squaring function and name it f. (Remember that we learned how to define our own functions in Chap. 5.) Then we use **NestList** which takes three arguments. The first is the function, in this case f. The second argument is the initial number, or point, in this case 2, and the last argument tells *Mathematica* how far to follow the orbit. In this case we have entered 5, telling *Mathematica* to stop after the function has been applied five times. **NestList** always includes the initial number as the first element of the list, so in this case we get a list with six elements.

It should be clear that if we start with any number larger than 1, then its orbit will go off to infinity. Of course if the number is really close to 1 it might take a lot more iterations of f before the orbit starts to get big. The next example repeats the first experiment but with an initial value of 1.000000000001.

Example 10.1.2

In[3]:=
```
(* following the orbit of 1.000000000001 *)
f[x_] := x^2
NestList[f, 1.000000000001, 50]
```

Out[4]= {1., 1., 1., 1., 1., 1., 1., 1., 1., 1., 1.,
1., 1., 1., 1., 1., 1., 1., 1., 1., 1.,
1., 1.00001, 1.00002, 1.00003, 1.00007,
1.00013, 1.00027, 1.00054, 1.00107, 1.00215,
1.0043, 1.00863, 1.01733, 1.03496, 1.07114,
1.14735, 1.3164, 1.73291, 3.00299, 9.01797,
81.3237, 6613.54, 4.3739×10^7, 1.9131×10^{15},
3.65994×10^{30}, 1.33952×10^{61}, 1.7943×10^{122},
3.21952×10^{244}, $1.036529965760730 \times 10^{489}$}

The first few elements of the orbit are so close to 1 that *Mathematica* just displays 1, but sure enough, by the time we get to the 50th iterate the number has gotten huge and clearly the orbit will head off to infinity.

If we start with an initial value of 1 however, then the orbit will not go anywhere! Since $1^2 = 1$ the orbit will just stay at the number 1. We say that 1 is a *fixed point* of f. Are there any other fixed points of f? We need to solve the equation $f(x) = x$ to find the fixed points of f. In this case we need to solve $x^2 = x$. This is pretty simple, but let's use *Mathematica* to find the solutions! The next example does this.

Example 10.1.3

In[5]:= (* finding the fixed points of f *)
        f[x_] := x^2
        Solve[f[x] == x, x]

Out[6]= {{x → 0}, {x → 1}}

We see that f has exactly two fixed points: 0 and 1. These are the only numbers that will not change when we square them. So each one has an unchanging orbit. What do the orbits of all other numbers do? We have seen that the numbers greater than 1 all behave similarly: their orbits grow without bound and head off to infinity. It should also be clear that numbers that lie between 0 and 1 have orbits that go to, or *converge* to, 0. If we start with .5 for example, **NestList** gives its orbit as:

Example 10.1.4

In[7]:= (* following the orbit of .5 *)
        f[x_] := x^2
        NestList[f, .5, 10]

Out[8]= {0.5, 0.25, 0.0625, 0.00390625,
        0.0000152588, 2.32831×10^{-10}, 5.42101×10^{-20},
        2.93874×10^{-39}, 8.63617×10^{-78},
        7.45834×10^{-155}, 5.56268464626800×10^{-309}}

After only a few iterations the orbit of .5 is incredibly close to 0. Of course it will never actually reach 0. Only 0 squared is 0, so to end at 0 we would have to start at 0. But the orbit of .5 does converge to 0; that is, it becomes arbitrarily close to 0 as we go out farther and farther in the orbit.

Finally, what about the negative numbers? As soon as we square a negative number it becomes positive and we have already worked out what the orbit of each positive number is. So we can summarize what happens to every point as follows:

- Numbers bigger than 1, or smaller than -1, have orbits that go to infinity.
- The number 1 is fixed. The number -1 goes to 1 where it remains.
- Nonzero numbers between -1 and 1 have orbits that converge to 0.
- Zero is fixed.

This example, as simple as it is, exhibits the kinds of orbits that are typical of many dynamical systems. We see points that there are fixed points (0 and 1) as well as a point that is *eventually* fixed, namely -1. (We'll say that a point is eventually fixed if its orbit eventually reaches a fixed point.) All the other orbits either converge to zero or to infinity.

The fixed points of this example are also quite interesting in their behavior. Zero is a fixed point and, moreover, all points that are really close to zero have orbits that converge to zero. For this reason we will call zero an *attracting* fixed point. On the other hand, even though 1 is fixed, points that are near to 1 have orbits that go away from 1 and this happens no matter how close we start to 1. So we call 1 a *repelling* fixed point. Finally, notice that lots of points have orbits that converge to the attracting fixed point 0. In fact, all points between -1 and 1 have orbits that converge to 0. So we call the set of all these points the *basin of attraction* of 0.

At this point we have a complete and total understanding of the *dynamics* of the function $f(x) = x^2$; that is to say, we understand completely the orbits of all points under f. In general, given any function f, this will be our goal. It turns out that fairly simple looking functions can have some incredibly complicated dynamics. In fact, if we just add a constant c to the present example and look at the function $f(x) = x^2 + c$, the dynamics can change considerably!

10.2 Graphical Analysis

If we look at orbits from a graphical point of view, it can really help us understand the dynamics of a function. With any function f it is always good to first identify the fixed points, if there are any. These are places where $f(x) = x$, or graphically, where the graph of $f(x)$ and the graph of x intersect. For example, suppose we consider the function $f(x) = \cos x$. Let's plot the graphs of both $\cos x$ and x and see where they intersect. From the next example we see that they intersect at exactly one point. Here we have used **Plot** to graph both functions simultaneously. We can see from the plot that the fixed point is between .6 and .8, but what is it exactly?

Example 10.2.1

```
In[20]:= (* the graphs of f[x] and x intersect
         at the fixed points of f[x] *)
    f[x_] := Cos[x]
    Plot[{f[x], x}, {x, -Pi, Pi},
     PlotStyle → {Red, {Blue, Dashing[.02]}},
     AspectRatio → Automatic]
```

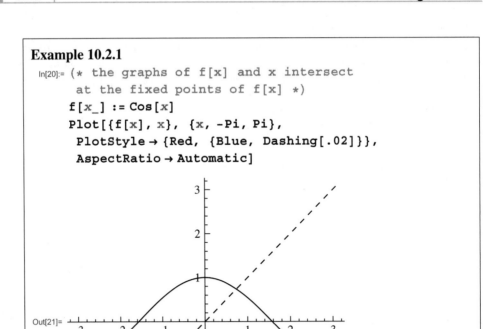

Out[21]=

Let's ask *Mathematica* to solve the equation $\cos x = x$ in order to find the fixed point. Using **FindRoot** we obtain the following. (You should try using **Solve** or **NSolve** to see what happens.)

Example 10.2.2

```
In[11]:= (* finding the fixed point of Cos[x] *)
    FindRoot[Cos[x] == x, {x, .5}]
Out[11]= {x → 0.739085}
```

After finding all the fixed points (only one in this case), a good thing to do next is to decide if the fixed point is repelling, or attracting, or neither. We can experiment with **NestList** to follow the orbits of some nearby points. This might provide some good evidence. Let's follow the orbit of a point close to the fixed point.

Example 10.2.3

In[12]:= (* following the orbit of a point near
       the fixed point of Cos[x] *)
    NestList[Cos, .739, 20]

Out[12]= {0.739, 0.739142, 0.739047, 0.739111, 0.739068,
       0.739097, 0.739077, 0.73909, 0.739082,
       0.739088, 0.739083, 0.739086, 0.739084,
       0.739086, 0.739085, 0.739085, 0.739085,
       0.739085, 0.739085, 0.739085, 0.739085}

Here we see that the point .739 has an orbit that apparently converges to the fixed point. In fact, after a while all the elements of the orbit are listed as .739085. So they have either landed on the fixed point exactly, or *Mathematica* is rounding off here and all these numbers are simply rounding off to the same thing. Looking at the beginning of the orbit we can also see that the elements of the orbit bounce back and forth, alternately smaller and then bigger than the fixed point. So, based on this one example, it certainly looks like the fixed point might be attracting.

We can make this completely clear by a graphical method known as Graphical Analysis. The idea is to plot the points of an orbit in such a way that we can see the order of the points and hence see where the orbit is heading. Consider Example 10.2.4.

We start by defining the function we want to study, in this case $\cos x$. Next we define the initial point x_0 of the orbit, the number n of iterations of f we want to use, and finally compute the orbit. We use **Plot** to plot the graphs of both $f(x)$ and x and we use **PlotStyle** to make the graph of f red and the graph x both blue and dashed. So far this should be familiar. But, after the plot has been rendered, we use **Epilog** to add in the arrows that you see in the plot. Before we see how *Mathematica* does this, let's understand what we are looking at.

The initial value in this example is .2. To find $f(.2)$ we need to draw a vertical line at $x = .2$ and see where it intersects the graph. The height of this intersection point is $f(.2)$. So what we have done is draw a vertical arrow from the point (.2, .2) up to the graph at (.2, $f(.2)$). We then draw a horizontal arrow from there over to the point ($f(.2)$, $f(.2)$). If the orbit is

$$x_0, x_1, x_2, x_3, \ldots$$

where each $x_{i+1} = f(x_i)$, then we have plotted the points (x_0, x_0), (x_1, x_1), (x_2, x_2), ... and connected each consecutive pair of points with arrows that go from (x_i, x_i) to $(x_i, f(x_i))$ and then from $(x_i, f(x_i))$ to (x_{i+1}, x_{i+1}). This pair of arrows is just what we need to "see" the action of f.

Example 10.2.4

```
In[148]:= (* basic cell for Graphical Analysis *)
       f[x_] := Cos[x] (* the function to study *)
       x0 = .2; (* initial point of orbit *)
       n = 3; (* number of iterations of f *)
       orbit = NestList[f, x0, n];
       Plot[{f[x], x}, {x, 0, Pi / 2},
        PlotStyle -> {Red, Blue},
        AspectRatio → Automatic,
        (* add arrows that follow orbit *)
        Epilog →
         Table[
          Arrow[{{orbit[[i]], orbit[[i]]},
            {orbit[[i]], orbit[[i + 1]]},
            {orbit[[i + 1]], orbit[[i + 1]]}}],
          {i, 1, n}]
       ]
```

Out[152]=

Now let's see how we use **Epilog** to draw in all the arrows. The **Table** function creates a list of **Arrow** graphics primitives that are then plotted with **Epilog**. Each **Arrow** function uses a pair of consecutive points from **orbit** to draw two consecutive arrows. **Arrow** is just like **Line**; if we enter a list of points **Arrow** will connect them in succession. We have entered three points into the **Arrow** function. The first and last points are on the graph of x while the middle point is on the graph of $f(x)$.

Example 10.2.4 is a cell that we can use over and over for lots of different functions. We just need to change the function, the initial point, and the number of iterations we want to plot. However, if we use a large value of n the arrowheads will tend to clutter up the plot and it may be better to forsake the arrowheads, using **Line** instead of **Arrow**.

From the example it is easy to see what the orbit is doing! It is converging to the fixed point, but in so doing, it alternates between being larger and smaller than the fixed point.

What we can also see from graphical analysis is that the nature of a fixed point p, that is, whether p is attracting or repelling, depends entirely on the slope of f at the fixed point! To see why the derivative $f'(p)$, which is the slope of the tangent line to f at p, determines the attracting or repelling nature of p, imagine that we have zoomed in really close on the intersection of the graphs of $f(x)$ and x; so close that the graph of $f(x)$ appears as a straight line with slope $f'(p)$. Now if we were to replace the graph of $f(x)$ with its tangent line at p we wouldn't be able to tell the difference! So the nature of the fixed point really only depends on the slope there. But how does it depend on the slope? Figure 10.1 shows two functions, each with a fixed point but with slopes of less than 1 and more than 1 at the fixed point. We have zoomed in so far that each function appears straight and hence is indistinguishable from its tangent line at the fixed point. In each case we follow the orbits of two points near the fixed point.

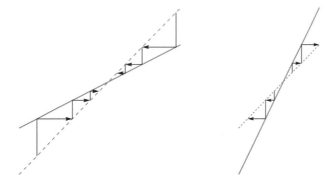

Figure 10.1   A fixed point p of $f(x)$ is attracting if $-1 < f'(p) < 1$ and repelling if $f'(p) < -1$ or $f'(p) > 1$.

Clearly if the slope of f at the fixed point is more than 1, then the fixed point is repelling, while if the slope of f at the fixed point is less than 1, then the fixed point is attracting. We summarize this in the following important theorem:

Theorem: If p is a fixed point of f, then

- if $|f'(p)| < 1$, then p is attracting,
- if $|f'(p)| > 1$, then p is repelling, and
- if $|f'(p)| = 1$, then p could be attracting, repelling, or neither.

To see that the last statement of the theorem is true requires three different examples where in each case the slope at the fixed point is 1 or -1 and yet we get fixed points of different natures. You will be asked to investigate such examples in the Quiz at the end of this chapter.

10.3 The Quadratic Family

We have already considered the function $f(x) = x^2$ and completely worked out what happens to the orbit of every point. But what if we add a constant c? What are the dynamics of the function $f(x) = x^2 + c$ where c is any real number? How does it depend on c? This is a perfect place to use **Manipulate**. Let's graph the function $x^2 + c$ with c as a parameter that we can manipulate. As we do this we can watch the graph move. At first, let's just focus on the fixed points and their nature. Example 10.3.1 shows the result, but of course, you need to do this yourself so that you can manipulate the parameter c. We have set it up so that the initial value of c is 0.5.

Example 10.3.1

```
In[15]:= (* watching the fixed points of f[x]=
         x² + c change with c *)
         Manipulate[
          Plot[{x^2 + c, x}, {x, -2, 2},
           AspectRatio → Automatic,
           PlotStyle → {Red, {Blue, Dashing[.02]}}
          ]
          , {{c, .5}, -2, .5}
         ]
```

Example 10.3.1 (Continued)

Out[15]=

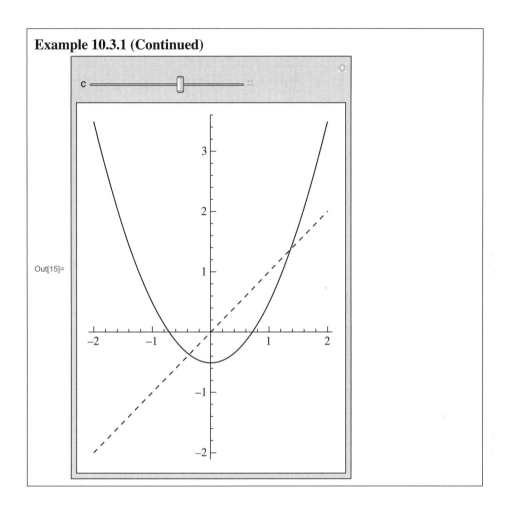

As we move the slider and vary c, the parabola moves up and down. Clearly if c is too large, then the parabola lies entirely above the graph of x and hence there are no fixed points. But if we lower the value of c the parabola will move down and eventually cross the graph of x. When they first become tangent there is exactly one fixed point, but as the parabola continues to move down the single fixed point splits into two fixed points. Let's find the values of the fixed points as functions of c. We can use **Solve** to do this.

Here we first used **Solve[x²+c==x, x]** to find the roots and then substitution to form a list of the two roots. We have named the roots p and q. Furthermore, we see that the roots are equal precisely when $c = .25$. This is when the parabola is tangent to the diagonal line $y = x$ and there is only one fixed point. If $c > .25$ then the parabola is entirely above the line $y = x$ and p and q become imaginary; there are no real fixed points. But if $c < .25$ then there are two real fixed points and $p < q$.

Example 10.3.2

In[22]:= (* finding the fixed points of $x^2 + c$ *)
$\{p, q\} = x /. Solve[x^2 + c == x, x]$

Out[22]= $\left\{ \frac{1}{2} \left(1 - \sqrt{1 - 4c}\right), \frac{1}{2} \left(1 + \sqrt{1 - 4c}\right) \right\}$

In[24]:= (* finding out when the fixed points
are equal *)
Solve[p == q, c]

Out[24]= $\left\{ \left\{ c \to \frac{1}{4} \right\} \right\}$

It is not hard to see that if $c > .25$, then every orbit goes off to infinity. You should experiment with Graphical Analysis to convince yourself of this. Just use Example 10.2.4 with f defined appropriately. So, the case with $c > .25$ is not that interesting. Much more interesting dynamics occur when $c \le .25$.

We can see something we haven't seen before if we let $c = -1$. Now $f(x) = x^2 - 1$. Let's see what the orbit of zero is. We have $f(0) = -1$ and $f(-1) = 1 - 1 = 0$. So the orbit of 0 is

$$0, 1, 0, 1, 0, 1, 0, 1, \ldots$$

The orbit just oscillates back and forth between 0 and 1. We call such an orbit *periodic* and say that its *period* is two since two iterations of the function bring us back to where we began. We'll also refer to such an orbit as a *2-cycle* and furthermore, we'll call both 0 and 1 *periodic points* or *period 2 points*.

If we let $c = -1.755$ something else interesting happens. Let's look at the orbit of 0 now. The results are displayed in Example 10.3.3.

Example 10.3.3

In[116]:= `f[x_] := x^2 - 1.755;`
`NestList[f, 0, 29]`

Out[117]= {0, -1.755, 1.32502, 0.000691251,
      -1.755, 1.32502, 0.000686806,
      -1.755, 1.32502, 0.000686863,
      -1.755, 1.32502, 0.000686862, -1.755,
      1.32502, 0.000686862, -1.755, 1.32502,
      0.000686862, -1.755, 1.32502, 0.000686862,
      -1.755, 1.32502, 0.000686862, -1.755,
      1.32502, 0.000686862, -1.755, 1.32502}

The orbit appears to be converging to a 3-cycle! Using our basic cell for Graphical Analysis (Example 10.2.4) gives the following output.

Example 10.3.4
```
(* using Graphical Analysis to follow the orbit of
0 when c = -1.755. *)
```

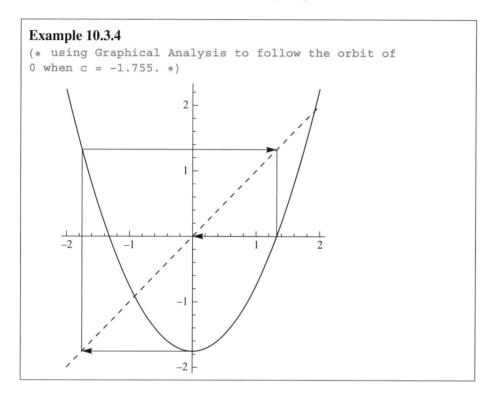

Note that the orbit of 0 is not exactly a 3-cycle itself. First of all, it does not return to 0 exactly. But even if we consider starting at -1.755, two iterations bring us to 0.000691251 and three more iterations bring us to 0.000686806 which is not quite the same. Still, after 29 iterations *Mathematica* displays the same three numbers (after rounding off) over and over.

It turns out that changing the parameter c can introduce periodic cycles of different length. It would be nice to methodically experiment with different values of c. Let's have *Mathematica* vary c automatically and for each value of c follow the orbit of zero. Since the orbit of zero might converge to a fixed point, or perhaps to a cycle of some period, it makes sense to follow the orbit for a few thousand iterations to see where it has ended up. But we certainly don't want to print out orbits that are thousands and thousands of iterations long! A nice thing to do would be to follow an orbit for say 1000 iterations, not printing any of these values, and then print out the next 10 or 20 iterations. We can do this if we make use of the **Nest** function, which is similar to **NestList**. The function **Nest**[*f*, *x*, *n*] will simply return the *n*th iteration of *x* under f, as opposed to all the iterates of *x* up to the *n*th one as **NestList** would do. Consider the following example.

Example 10.3.5

```
In[27]:= (* following the orbit of 0,
   ignoring the first 9999 iterates *)
   f[x_] := x^2 - .6;
   NestList[f, Nest[f, 0, 10 000], 5]

Out[28]= {-0.421954, -0.421954, -0.421954,
   -0.421954, -0.421954, -0.421954}
```

First we define the function as usual, with $c = -.6$. Then we use **NestList** to print out an orbit with 5 iterations. But we are not printing the orbit of 0, instead we are starting at **Nest[f, 0, 10000]** which is the 10000-th iteration of 0 under f. So what this does is skip the first 9999 iterations of the orbit of 0 and then print out the next 6 elements of the orbit. In this case, with $c = -.6$, the orbit of 0 seems to have converged to the fixed point -0.421954.

What we'd like to do now is repeat this experiment but with different values of c and have *Mathematica* automatically vary c. We do this in Example 10.3.6, with c

---

Example 10.3.6

```
In[9]:= (* examining the fate of 0 for
   different values of c *)
   f[x_] := x^2 + c;
   Do[
    tailOfOrbit =
     NestList[f, Nest[f, 0, 10 000], 6];
    Print["c=", c, " ", tailOfOrbit],
    {c, .25, 0, -.05}
   ]
c=0.25 {0.4999, 0.4999, 0.4999,
   0.4999, 0.4999, 0.4999, 0.4999}
c=0.2 {0.276393, 0.276393, 0.276393,
   0.276393, 0.276393, 0.276393, 0.276393}
c=0.15 {0.183772, 0.183772, 0.183772,
   0.183772, 0.183772, 0.183772, 0.183772}
c=0.1 {0.112702, 0.112702, 0.112702,
   0.112702, 0.112702, 0.112702, 0.112702}
c=0.05 {0.0527864, 0.0527864, 0.0527864,
   0.0527864, 0.0527864, 0.0527864, 0.0527864}
c=0. {0., 0., 0., 0., 0., 0., 0.}
```

varying from $c = .25$ down to $c = 0$ in steps of .05. For each value of c, we follow the orbit of 0 for 10000 iterations and then print out 6 more iterations. When we print, we print the value of c and then the 6 elements of the orbit.

We have done this by using a **Do** loop. The loop is executed six times, once for each value of c in the set $\{.25, .20, .15, .10, .05, 0\}$. Notice that we define the function f before the loop and yet c is changing with each pass through the loop. Since we defined f using **SetDelayed** as opposed to **Set** (i.e., := instead of =), the correct definition of the function is used. If we had used **Set** instead, it would not work properly.

Let's make sure we understand the body of the **Do** loop. The first line defines **tailOfOrbit** as iterates 10000 through 10006 by using both **NestList** and **Nest** as previously described. Finally, we use the **Print** function to print out the value of c and the tail end of the orbit of 0. The **Print** function can take any number of arguments separated by commas. Each argument is something that will be printed and in this case is either a text string (any characters delimited by quotes), or a variable.

Our experiment reveals that for these six values of c the orbit of zero seems to be converging to a fixed point. But that fixed point is changing with c. In fact, the orbit of zero is converging to the fixed point $p = (1 - \sqrt{1 - 4c})/2$ which we computed earlier. Apparently this is an attracting fixed point and 0 is in its basin of attraction. You should convince yourself of this by using Graphical Analysis to follow the orbit of zero for these values of c.

What happens as we lower c even more? The next example repeats the experiment but with a different range of c values.

Look carefully at the data! Something quite remarkable happens as c passes through the value $-.75$. When $c > -.75$ the orbit of zero is converging to an attracting fixed point. But when $c < -.75$ the orbit of zero now appears to be converging to a 2-cycle! It turns out that at $c = -.75$ the attracting fixed point p transitions from attracting to repelling. Remember our earlier discussion on the nature of a fixed point as determined by the slope of the function there? Well, when $f(x) = x^2 - .75$ the fixed point p is equal to $-.5$ and the slope of f at p equals -1 since $f'(x) = 2x$. When c is above $-.75$ the absolute value of the slope of f at c is less than 1. But when $c < -.75$, the absolute value of the slope of f at p is more than 1. At $c = -.75$, the fixed point p turns from attracting to repelling. You should go back to Example 10.2.4 (and set $f(x) = x^2 - .75$) and use Graphical Analysis to watch this happen! Repeat the experiment for various values of c just on either side of $-.75$. It is easy to see the slope of f at p changing as we vary c. Even more exciting is the fact that at the very moment p changes from attracting to repelling, a 2-cycle is born and the orbit of 0 is now attracted to this 2-cycle! Since we may think of a fixed point as a periodic point of period 1, this change from a period 1 attractor to a period 2 attractor is called a *period doubling bifurcation*. We can see what is going on if we consider the graphs of both $f(x)$ and $f(f(x))$.

Example 10.3.7

```
(* examining the fate of 0 for
  different values of c *)
f[x_] := x^2 + c;
Do[
 tailOfOrbit =
  NestList[f, Nest[f, 0, 10000], 6];
 Print["c=", c, " ", tailOfOrbit],
 {c, -.7, -.8, -.025}
]
c=-0.7 {-0.474679, -0.474679, -0.474679,
  -0.474679, -0.474679, -0.474679, -0.474679}
c=-0.725 {-0.487421, -0.487421, -0.487421,
  -0.487421, -0.487421, -0.487421, -0.487421}
c=-0.75 {-0.492909, -0.50704, -0.49291,
  -0.50704, -0.492911, -0.507039, -0.492911}
c=-0.775 {-0.341886, -0.658114, -0.341886,
  -0.658114, -0.341886, -0.658114, -0.341886}
c=-0.8 {-0.276393, -0.723607, -0.276393,
  -0.723607, -0.276393, -0.723607, -0.276393}
```

Here is an important idea: If P is a period 2 point of $f(x)$, then P is brought back to itself after two iterations of f and so P is a fixed point of $f(f(x))$. So if we find the fixed points of the function $f(f(x))$ we will be finding 2-cylces of $f(x)$. Or will we? Notice that if a point is fixed by $f(x)$ then it is fixed by $f(f(x))$. What if the only fixed points of $f(f(x))$ are just the fixed points of $f(x)$ itself? This can happen and in this case there simply are no period 2 points of f. But, it might be the case that $f(f(x))$ has fixed points that are not fixed by $f(x)$. These will be genuine period 2 points of f. Let's use *Mathematica* to simultaneously graph both $f(x)$ and $f(f(x))$. The result is pictured in Example 10.3.8, with **Manipulate** used so that we can vary c.

The graph of $f(x)$ is the parabola. The graph of $f(f(x))$ is the 4-th degree polynomial that is drawn slightly thicker. The graphs are pictured with $c = -.5$. Notice that both $f(x)$ and $f(f(x))$ have the same fixed points as they both intersect the diagonal line $y = x$ at the same two points, p and q, where p is to the left of q. But what happens as we lower c? The graph of f will move down. What will happen to the graph of $f(f(x))$? You should *definitely* be trying this out yourself!

Example 10.3.8

```
In[14]:= (* graphing both f(x) and f(f(x)) *)
Manipulate[
 f[x_] := x^2 + c;
 Plot[{f[x], f[f[x]], x}, {x, -2, 2},
   PlotRange -> {{-2, 2}, {-2, 2}},
   AspectRatio -> Automatic,
   PlotStyle -> {
     {Red, Thickness[.002]},
     {Red, Thickness[.005]},
     {Blue, Dashing[.02]}
    }
 ]
 , {c, -2, .25}
 ]
```

Out[14]=

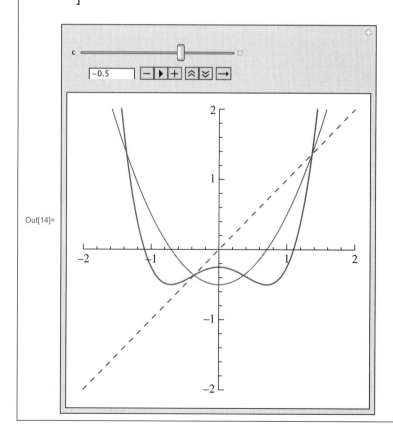

Let's first think about the slope of $f(x)$ at the two fixed points p and q. In the following Example 10.3.9 we use **Reduce** to find out when the slope of f at the fixed points has absolute value 1 or less. Notice the option **Reals** that we have used with **Reduce** so that *Mathematica* knows that c is a real parameter. We see that the slope at q is only ± 1 when $c = .25$. In fact it is obvious from the graph of $f(x)$ that the slope at q is 1 precisely when $c = .25$ and more than 1 when $c < .25$. So for $c < .25$, q is always a repelling fixed point.

Example 10.3.9

```
In[15]:= (* where are the fixed points attracting? *)
         f[x_] := x^2 + c
         {p, q} = x /. Solve[f[x] == x, x];
         Reduce[Abs[f'[p]] <= 1, c, Reals]
         Reduce[Abs[f'[q]] <= 1, c, Reals]
```

$$\text{Out[17]}= \quad -\frac{3}{4} \leq c \leq \frac{1}{4}$$

$$\text{Out[18]}= \quad c == \frac{1}{4}$$

On the other hand, when $-.75 < c < .25$ the slope of $f(x)$ at p is between -1 and 1 and so p is an attracting fixed point in this range of c. Looking again at Example 10.3.8, it should be clear that the slope of $f(x)$ at p is negative and getting steeper as c decreases. The moment that $c = -.75$ the slope of $f(x)$ at p hits -1. Thus at this moment, when $c = -.75$, the fixed point p turns from attracting to repelling.

Now let's turn our attention to the graph of $f(f(x))$. As you play with the slider controlling c in Example 10.3.8 you see that the graph of $f(f(x))$ has only two fixed points if $c \geq -.75$. However, as we lower c, the local maxima of $f(f(x))$ at zero moves up and the two local minima of $f(f(x))$ move down. When we reach $c = -.75$ the slope of $f(f(x))$ at p is exactly 1! When $c < -.75$ the graph of $f(f(x))$ now intersects the diagonal line $y = x$ in four points: the two fixed points p and q of f, but also two new points. These new points are a 2-cycle. They are on opposite sides of p and if we raise c back up again to $c = -.75$ the two points of the 2-cycle come together at p. Note also that when this 2-cycle is born, the slope of $f(f(x))$ at each point of the 2-cycle is clearly less than 1. Hence these are both attracting fixed points of $f(f(x))$ or equivalently, they form an attracting 2-cycle of f. This explains how the attracting fixed point p of f becomes repelling and simultaneously gives rise to an attracting 2-cycle as c passes through the value $-.75$.

Perhaps even more exciting is the fact that as c goes even lower the attracting 2-cycle becomes repelling while an attracting 4-cycle is born! See if you can experimentally find the value of c where this occurs! In fact these period doubling bifurcations continue with the creation of an 8-cycle and then a 16-cycle and on and on forever!

We can attempt to capture all of this discussion into a single *orbit diagram* which is shown in Example 10.3.10. Let's understand the diagram and then see the *Mathematica* code that was used to plot it. The horizontal axis represents the constant c and ranges from -2 to $.25$. For a fixed value of c imagine following the orbit of zero until it settles down, perhaps at (or, more accurately, near) an attracting fixed point, or an attracting 2-cycle, and so on. We then plot that point, or points if the orbit is attracted to a cycle, on the vertical scale. For example, we have seen that if $c = .25$ then there is a single attracting fixed point at $x = .5$. Thus the point $(.25, .5)$ is plotted in the orbit diagram. We have also seen that as we lower c toward zero, the fixed point p remains attracting but its value drops toward zero. Thus the orbit of zero still converges to p, but p itself is getting smaller. This continues until c reaches $-.75$ where the period doubling bifurcation occurs. Hence we see a single curve in the orbit diagram that starts out at the point $(.25, .4)$ and drops as we move left to $c = -.75$. Notice in Example 10.3.10 that this curve appears to pass through the point $(0, 0)$. This makes sense: when $c = 0$, $p = 0$ is an attracting fixed point. When we reach $c = -.75$, the attracting fixed point becomes repelling and the attracting 2-cycle is born. Thus our orbit diagram splits into two

Example 10.3.10

(* the orbit diagram for the family f(x) = x² + c *)

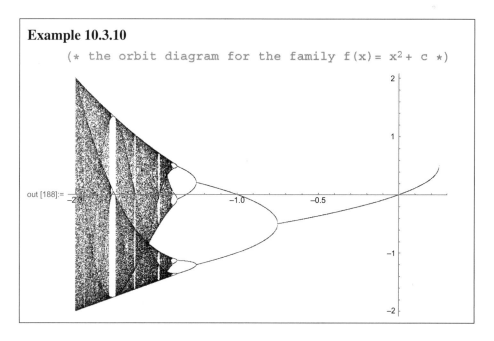

out [188]:=

Example 10.3.10 (Continued)

```
In[23]:= (* orbit diagram plot *)
    f[x_] := x^2 + c; (* define f *)
    points = {}; (* start with empty list *)
    Do[
      (* find tail of orbit *)
      tail = NestList[f, Nest[f, 0, 10 000], 100];
      (* convert points in orbit to points
       in plane *)
      newPoints = {c, #} & /@ tail;
      (* add points to cumulative list *)
      points = Join[points, newPoints]
      , {c, .25, -2, -.001}
    ]
    (* plot all the points in diagram *)
    ListPlot[points,
      PlotStyle → {{Blue, PointSize[.0001]}}
    ]
```

curves. We have seen in Example 10.3.7 that when $c = -.8$ the attracting 2-cycle is $\{-.723607, -.276393\}$. Thus the points $(-.8, -.723607)$ and $(-.8, -276393)$ are plotted in the orbit diagram.

The code used to produce the orbit diagram is given above. Let's see how it works. There are four instructions: the first defines the function and the second defines the list **points**, which starts out empty. The third is a **Do** loop (to compute the points of the plot) and the fourth is a **ListPlot** that plots the points.

The set of points that we are going to plot at the end with the **ListPlot** function is named **points**. The first line defines this variable as a list with no elements. All the work is in the loop, which is indexed by the parameter c. We let c start at .25 and go down to -2 in steps of .001. (The first time you run this, choose a smaller stepsize so that it will not take so long!) For each value of c we execute all the steps in the body of the loop. First we compute the tail of the orbit, this time ignoring the first 10000 iterations and then saving the next 100 iterations. (If the orbit has fallen into a 2-cycle then this is highly redundant: we are saving the same two points 50 times each! But if the orbit has fallen into a 123-cycle for example, we will not even be recording the entire cycle.) What we want to do with these elements of the orbit is plot them on the vertical line $x = c$, so we need to take each point and convert it into a point in the plane with first coordinate c. We do this by mapping the pure function **(c, #)&** onto **tail**. Having found all the points in this vertical slice of the orbit diagram, we now want to save them in our big list of points. We do this with the **Join** function which is used to join two or more lists together into a single list. We replace the list **points** with this new list which is made of the old **points** list together with the **newPoints** that were generated on this pass through the loop.

The orbit diagram shown in Example 10.3.10 is a truly amazing plot. As we go from right to left we can see the period doubling bifurcations from 1 to 2, 2 to 4, and can almost make out the 4 to 8 doubling. The 8 to 16 doubling is too tiny to see, but if we rerun Example 10.3.10 with c only varying from $c = -1.392$ down to $c = -1.398$ we can bring it into view. We can also see in Example 10.3.10 values of c that give rise to 3-cycles, or 5-cycles. It is interesting to zoom in on these regions by rerunning Example 10.3.10 with the appropriate range of c values. We will not delve deeper into the complexities of the orbit diagram, but encourage you to explore it more yourself!

10.4 Julia Sets

So far we have considered the quadratic family of functions $f(x) = x^2 + c$, but only with real numbers. Why not allow complex numbers? If x is any complex number we can still square it and add c. What happens if we expand our horizons to complex numbers?

In mathematics it is common to use the letter z for a complex variable and to additionally write $z = x + iy$ so that x and y represent the real and complex parts of the variable. Here, of course, i is the imaginary number $i = \sqrt{-1}$. In order to understand the orbits of different points under the function $f(z) = z^2 + c$, we need a couple of important facts about complex numbers. Given a complex number $x = x + iy$ its *magnitude*, or *absolute value*, is given by $|z| = \sqrt{z^2 + y^2}$, and can be computed in *Mathematica* by using the absolute value function **Abs**. Graphically, the absolute value of a complex number is simply its distance from the origin. An extremely important fact about complex numbers is that the absolute value of the product of two numbers is the product of their absolute values. That is

$$|zw| = |z||w|$$

Another important feature of a complex number z is its *argument*. This is the angle made by the positive real axis (the x-axis) and the ray made by connecting the origin to z. In *Mathematica* we can use **Arg[z]** to compute the argument of z. If $z = x + iy$ then $\tan(\arg z) = x/y$. Another important property of complex number is that the argument of the product of two numbers is the sum of the arguments of the numbers. That is

$$\arg(zw) = \arg z + \arg w$$

Let's start with $c = 0$ and use the above facts to understand the dynamics of $f(z) = z^2$. Clearly 0 and 1 are the only fixed points, just as before when we considered only real numbers. Because squaring a number will square its absolute value, we see that all complex numbers with absolute value less than 1 will have orbits that converge to zero. This is the set of all numbers inside the unit circle centered at the origin. Furthermore, all numbers located outside the unit circle have absolute values greater than 1 so their orbits will go off to infinity. What about a number with absolute value equal to 1? All of its iterates will continue to have absolute values of 1, so its orbit will stay on the unit circle. However, since arguments add when we multiply, the orbit will move round and round the unit circle. In fact if we only consider the *unit* complex numbers (the ones with absolute values of 1, and hence lying on the unit circle), the dynamics of f are incredibly complicated!

In general, the *Julia Set* associated to $f(z) = z^2 + c$ is the boundary of the set of points whose orbits go off to infinity. Thus with $c = 0$ the points whose orbits go off to infinity are the points that lie outside the unit circle and the Julia Set is the unit circle. How does the Julia Set change if we change c and how can we use *Mathematica* to draw the Julia Set? What we need to do, for a given value of c, is to start following the orbits of various points and see which orbits go off to infinity. This is not that easy since we can only follow an orbit for a few hundred or perhaps a few thousand iterations. If we stop following an orbit, how will we know if it is destined to go off to infinity or not? The key ingredient for us is the following theorem:

Theorem: *If $|c| \leq 2$ and any element in the orbit of z_0 under the function $f(z) = z^2 + c$ has an absolute value greater than 2, then the orbit of z_0 will go to infinity.*[1]

So, assuming that $|c| \leq 2$, we don't even need to follow the orbits of points that lie outside the circle of radius 2 centered at the origin. They already have absolute values greater than 2, so their orbits go to infinity. On the other hand, if a point is on or inside the circle of radius 2 we need to follow its orbit to see what it does. If after following it for a while, the orbit moves outside the circle of radius 2, then we can stop following it knowing that it is headed for infinity. The problem is with orbits that we follow for a while and which continue to stay on or inside the circle of radius 2. Such an orbit either never goes to infinity, or it does but we need to follow it further to find out.

The following example defines a function called *orbitLength* which will follow an orbit for up to 500 iterations but will stop short of that if the orbit ever goes outside the circle of radius 2. The function then returns the length of the orbit at the point when it stopped. Thus if the function returns a number less than 500 we know that the orbit went outside the circle of radius 2 and hence is headed for infinity.

---

[1] We will not prove this fact here, but it can be found on page 86 of *Chaos, Fractals, and Dynamics* by Robert L. Devaney, Addison-Wesley, 1990.

Example 10.4.1

```
In[101]:= (* follows the orbit of z0 under f(z) =
            z^2+c until orbit exceeds 2 in absolute
             value OR we hit 500 iterations *)
       orbitLength[z0_, c_] := Module[
         { z = N[z0], cc = N[c], steps = 0},
         While[steps < 500 && Abs[z] < 2,
          z = z * z + cc;
          steps++
         ];
         steps
        ]
```

Let's see how **orbitLength** works. To begin with, **orbitLength** takes two arguments, $z0$ and c. We are going to follow the orbit of $z0$ under the function $f(z) = z^2 + c$. Next, we use the **Module** function to define **orbitLength**. The first argument to the **Module** function is the list of local variables and in this case we define three: cc, z, and *steps*, all of which are initialized when they are declared. We really want $z0$ and c to be decimal approximations rather than exact numbers because dealing with exact numbers will be slower. So we use the numerical function **N** to immediately convert $z0$ and c to z and cc, respectively. We do this only so that we do not have to always remember to input decimal approximations to **orbitLength**; if we forget, the function will take care of it for us! The third local variable is called *steps* and will be used to count the number of iterations we take as we follow the orbit. After the declaration of the local variables comes the instructions that comprise the definition of **orbitLength**. There are only two: a **While** statement and then the output of the function, the variable *steps*. The **While** function provides us with a way to construct *indefinite* loops as opposed to the *definite* loop constructed with the **Do** function. The syntax of the **While** function is **While**[*test, body*] and the function works by repeatedly evaluating *test* and then *body* so long as *test* evaluates to true. In our case what we want to do is follow the orbit until either the orbit moves outside the circle of radius 2 or we iterate $f(z)$ 500 times. Our test is

$$\textbf{steps} < \textbf{500 \&\& Abs[z]} < \textbf{2}$$

We'll see why this does what we want in just a moment. Before doing that, let's move on to the simpler body of the loop which just consists of the two statements

$$\textbf{z=z*z+cc;}$$

$$\textbf{steps++}$$

Thus what we are doing in the body of the loop is simply moving z to the next point in the orbit and incrementing **steps** by 1. Going back to the test part of the loop, remember that we want to stop if the orbit moves outside the circle of radius 2. This explains half of the test condition, namely, **Abs[z]<2**. So long as this is true we want to keep on evaluating the body of the loop, that is, continue to follow the orbit. But we want to stop if we reach the 500-th iterate. This is where **steps** comes in. We start out with **steps** equal to 0 and we increment **steps** by 1 each time we evaluate the body of the loop. This is accomplished by using the **Increment** function **++**. In general, if **a** is any variable in *Mathematica* then **a++** will increment **a** by one. Finally, notice that this truly is an indefinite loop. The number of times the body of the loop is repeated is not fixed or predetermined. Instead it depends on the input $z0$ and c.

To construct the Julia Set for $f(z) = z^2 + c$ we simply need to test every point in the complex plane with **orbitLength** to see if its orbit goes off to infinity or not. But wait a minute! There are infinitely many points in the complex plane. How can we test them all? Obviously we can't. What we'll do instead is pick a rectangular portion of the complex plane and divide it by horizontal and vertical lines into a collection of smaller rectangles. Then we'll pick one point in each rectangle, say the lower left corner, and follow the orbit of that point. This corresponds quite nicely to the physical reality of our computer screen which is divided in exactly the same way into tiny *picture elements* or *pixels*. The next example defines the function **JuliaData** that will do just that.

Example 10.4.2

```
In[189]:= (* This function will test each point
       on a grid in the given region and return
       the orbit length for each point. *)
    JuliaData[c_,
       {{xmin_, xmax_}, {ymin_, ymax_}},
       xdiv_, ydiv_] :=
    Table[orbitLength[x + I y, c],
       {y, ymax, ymin, -(ymax - ymin) / ydiv},
       {x, xmin, xmax, (xmax - xmin) / xdiv}]
```

There are four arguments for the function **JuliaData**. The first is the constant c which defines the function $f(z) = z^2 + c$ which we are studying. Next is the list $\{\{xmin, xmax\}, \{ymin, ymax\}\}$ which defines the region of the complex plane that we want to study, namely the rectangle that extends horizontally from $xmin$ to $xmax$ and vertically from $ymin$ to $ymax$. The last two arguments are $xdiv$ and $ydiv$ which are the number of horizontal and vertical divisions of the rectangle. Finally, we define **JuliaData** by simply using the **Table** function to return the **orbitLength** at each point of the grid within the given rectangle. Since we are using two indices in the **Table** function, we create a list of lists, or in other words, a matrix. Notice that if $xdiv$ and $ydiv$ are both 1000, then there will be $1001 \times 1001 = 1,002,001$ points in the grid. **JuliaData** is going to return a very large matrix! Furthermore, for each of these points we are going to follow its orbit for up to 500 iterations. The total number of times that we might compute $f(z)$ is about 500 million! This computation might take a while!

Notice that the **Table** function in **JuliaData** has two indices with the y index first, counting down from $ymax$ to $ymin$, and the x index second, counting up from $xmin$ to $xmax$. Ordering the indices this way makes the orientation of the entries in the matrix match their orientation in the plane. That is, the data for the lower edge of the rectangle is in the last row of the matrix, and the data for the left edge of the rectangle is in the first column of the matrix. (Experiment with **Table[{i,j}, {i,0,2},{j,0,2}]** to see how *Mathematica* orders the table entries.)

The following example shows the output of **JuliaData** with only 10 divisions both vertically and horizontally. We have used the function **MatrixForm** to display the data as a matrix corresponding to the grid of points in the plane whose orbits we followed.

Example 10.4.3

In[116]:= **MatrixForm[**
       JuliaData[0, {{-2, 2}, {-2, 2}}, 10, 10]
      ]

Out[116]//MatrixForm=

$$
\begin{pmatrix}
0 & 0 & 0 & 0 & 0 & 0 & 0 & 0 & 0 & 0 & 0 \\
0 & 0 & 0 & 1 & 1 & 1 & 1 & 1 & 0 & 0 & 0 \\
0 & 0 & 1 & 1 & 2 & 2 & 2 & 1 & 1 & 0 & 0 \\
0 & 1 & 1 & 3 & 500 & 500 & 500 & 3 & 1 & 1 & 0 \\
0 & 1 & 2 & 500 & 500 & 500 & 500 & 500 & 2 & 1 & 0 \\
0 & 1 & 2 & 500 & 500 & 500 & 500 & 500 & 2 & 1 & 0 \\
0 & 1 & 2 & 500 & 500 & 500 & 500 & 500 & 2 & 1 & 0 \\
0 & 1 & 1 & 3 & 500 & 500 & 500 & 3 & 1 & 1 & 0 \\
0 & 0 & 1 & 1 & 2 & 2 & 2 & 1 & 1 & 0 & 0 \\
0 & 0 & 0 & 1 & 1 & 1 & 1 & 1 & 0 & 0 & 0 \\
0 & 0 & 0 & 0 & 0 & 0 & 0 & 0 & 0 & 0 & 0
\end{pmatrix}
$$

Looking at the data, we can see that points near the center of the rectangle have orbits that have not gotten larger than 2 in absolute value after 500 iterations. These orbits may or may not be headed for infinity. But around the perimeter of the rectangle we see lots of points whose orbits have gotten bigger than 2 in absolute value after only 0, 1, 2, or 3 iterations. The points that are labeled with zeroes are ones that were already outside the circle of radius 2 to begin with; we didn't follow their orbits at all. In this example we have $c = 0$, so what we are getting fits perfectly with what we expected. We know the set of points whose orbits go to infinity are all points outside the unit circle.

We need a way to visualize the output of **JuliaData** graphically. Fortunately, *Mathematica* has just what we need, the **ArrayPlot** function. Look what **ArrayPlot** does with this data in the next example. Each point of the grid where the orbit length is 500 is colored black and each point whose orbit escapes to infinity is colored white. In fact, **ArrayPlot** will use shades of gray for array values between the minimum, which it colors white, and the maximum, which it colors black, but in this case the gray level for a value of 1 versus 2 is so close that we cannot tell by looking at the graphic. As usual, **ArrayPlot** accepts all the usual options plus some that are special to it. You should definitely take a look at the Help Files to see what the possibilities are.

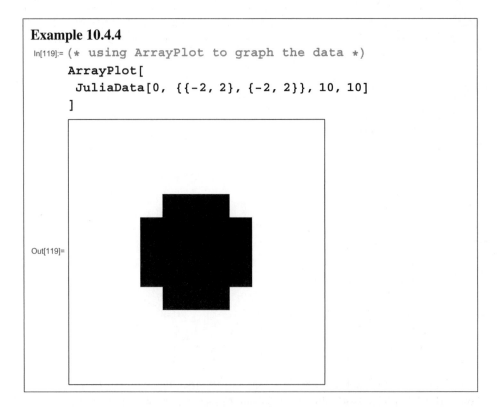

Example 10.4.4

```
In[119]:= (* using ArrayPlot to graph the data *)
          ArrayPlot[
            JuliaData[0, {{-2, 2}, {-2, 2}}, 10, 10]
          ]

Out[119]=
```

We have all the tools we need now to really investigate the Julia Sets of $f(z) = z^2 + c$ for lots of different values of c. It turns out that some pretty amazing sets arise once we start varying c. (Look ahead at the pictures!) But unfortunately, it can take a *really* long time to plot a Julia Set. Lucky for us though, *Mathematica* can be sped up considerably under certain circumstances. We need to make **orbitLength**, the main time-consuming part of our calculation, run faster and we can do this by *compiling* it. In general, *Mathematica* is not as fast as it could be because its functions can accept all sorts of types of input: integers, real numbers, complex numbers, lists of such numbers, matrices, variables, and so on. By telling *Mathematica* specifically what to expect as input we can greatly increase the speed of execution, often by a factor of 10. The following example shows the compiled version of **orbitLength**. It is nearly the same as the original **orbitLength**, except that we have wrapped it with the **Compile** function and changed the way that we have indicated the arguments $z0$ and c, telling *Mathematica* specifically to expect complex numbers. Since this is the only place in this book that we will be using compiled function we won't say more about compiling. You can find out more in the Help Files. You should try plotting a Julia Set with both the uncompiled and compiled version of **orbitLength** to see what a difference it makes. Use the **Timing** function to measure how long each computation takes.

Example 10.4.5

```
In[41]:= (* compiled version of orbitLength *)
    orbitLength = Compile[
        {{z0, _Complex}, {c, _Complex}},
        Module[
            { z = N[z0], cc = N[c], steps = 0},
            While[steps < 500 && Abs[z] < 2,
                z = z * z + cc;
                steps++
            ];
            steps
        ]
    ];
```

We close this section with a few examples of Julia Sets for different values of c. In Example 10.4.6 we have set $c = 0.360824 + 0.100376i$ and divided the square into 500 divisions both vertically and horizontally. Whoa! What a complicated picture, and yet, it has features that seem to repeat over and over on an ever smaller scale. We could "zoom" in by repeating the experiment with a smaller choice of original rectangle.

In Example 10.4.7 we change c to $0.64i$ and get a picture that again seems to have repeated features but is otherwise quite different from Example 10.4.6. Try experimenting with different values of c yourself!

Example 10.4.6

```
In[191]:= (* plotting a Julia Set *)
        data = JuliaData[.360824 + .100376 I,
          {{-2, 2}, {-2, 2}}, 500, 500];
        ArrayPlot[data,
          DataRange → {{-2, 2}, {-2, 2}},
          FrameTicks → Automatic
        ]
```

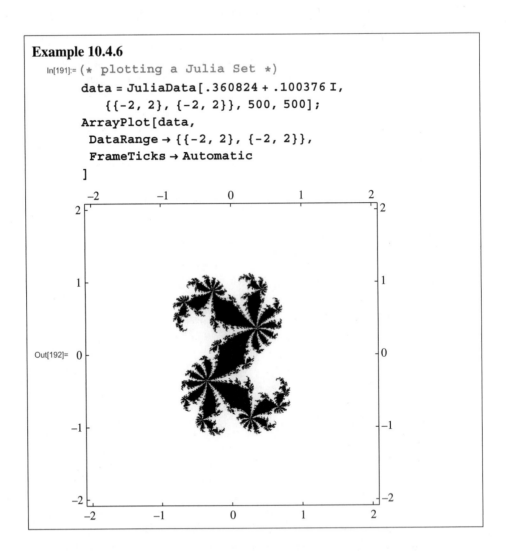

Notice the use of the option **DataRange** in the **ArrayPlot** function. Using this option, with a data range that matches the original coordinates of the rectangle, causes tick marks on the graphics frame to correspond with the actual x- and y-coordinates in the plane.

Example 10.4.7

```
In[195]:= (* plotting a Julia Set *)
       data = JuliaData[ .64 I, {{-2, 2}, {-2, 2}},
          500, 500];
       ArrayPlot[data,
        DataRange → {{-2, 2}, {-2, 2}},
        FrameTicks → Automatic
       ]
```

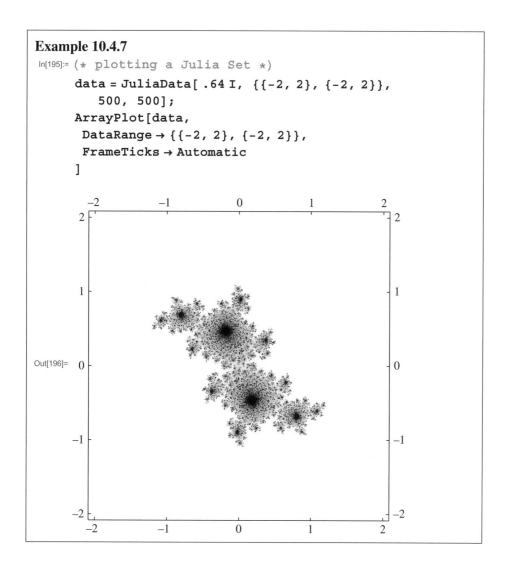

Out[196]=

10.5 Custom Coloring

The default coloring scheme in **ArrayPlot** is with a gray scale ranging from white at the minimum values in the array to black at the maximum values. But we can override this and provide various coloring schemes. Since this book is printed in black and white you will really need to try these examples yourself! We'll also look at a neat black and white coloring.

A nice coloring scheme can be obtained using the function **Hue**[h,s,b], which produces a color corresponding to h with a saturation of s and a brightness of b.

Each of the three parameters should be between 0 and 1, and as h goes from 0 to 1 the color runs through red, yellow, green, cyan, blue, magenta, and back to red again. Values of s and b outside the range of 0 to 1 are clipped while values of h outside this range are treated cyclically. **ArrayPlot** takes the data that we are trying to plot and rescales all the numbers so as to lie between 0 and 1, with the minimum value going to 0 and the maximum going to 1. So this is ideally suited to using the various color functions like **RGBColor, GrayLevel**, or **Hue** which all take arguments in the 0 to 1 range.

Example 10.5.1

```
In[30]:= (* drawing a Julia Set with custom
    coloring *)
color1 = If[# == 1.0, Black, Hue[50 #, 1, 1]] &;
data1 = JuliaData[
    -1, {{-2, 2}, {-2, 2}}, 500, 500
  ];
ArrayPlot[data1,
  ColorFunction -> color1
]
```

Out[32]=

In Example 10.5.1 we plot the Julia Set with $c = -1$ with a custom color function that we name **color1**. We want to color the points whose orbits go the full 500 iterations black, so if the input to **color1** is 1.0 we produce an output of black. Don't forget that **ArrayPlot** rescales the data so that the maximum value of 500 will be scaled to 1.0. For the shorter orbits which do not last to 500, we color them according to how long they are. It turns out that in a typical Julia Set calculation, most of the orbit lengths which are smaller than 500 are in fact much smaller. So we have multiplied the input by 50 to "smear out" the whole color range over each "decade" of orbit lengths. The next step is to include the option **ColorFunction→color1** in the **ArrayPlot** function. The previous example illustrates this with the Julia Set given by $c = -1$. The results in black and white are nowhere near as stunning as in color. You need to try this out yourself.

There is simply no end to possible coloring schemes and we invite you to try the following examples:

- **color2=If[#==1.0, Black, RGBColor[1-#, 100 #, 30 #1]]&**
- **color3=If[#==1.0, Black, RGBColor[0, 0, 100 #]]&**
- **color4=If[Mod[Floor[500 #], 2]==0, Black, White]&**
- **color5=If[#==1.0, Black, GrayLevel[30#]]&**

10.6 Making Movies

Changing the constant c in the family of functions $f(z) = z^2 + c$ has a huge effect on the Julia Set. Wouldn't it be cool to make an animation that shows the Julia Set changing as c changes? We could try to use **Manipulate** but the results will not be very good because of the long calculation times needed to plot a Julia Set. Instead we need to generate the successive frames of the movie and then view them in rapid succession as an animation. We can do this with **ListAnimate**. Let's see how to do this.

First, we need to generate the frames of the movie. Example 10.6.1 does this and displays all the frames.

In the first cell we make all the successive frames of the movie and place them in a list by using the **Table** function. Our movie will show the Julia Set as the constant c moves from 0 to $.75i$. We simply use c as the index for the **Table** function with a stepsize of $.05i$. Of course, to make a "finer" movie that runs "smoother" between frames we would use a smaller stepsize.

The second cell shows all the frames of the movie at once using the **Graphics-Grid** function which takes as its argument an array of graphics objects. So in order to use **GraphicsGrid** we need to take the frames that we generated and arrange them in an array. The function **Partition**[*list, k*] will do this by taking the elements

Example 10.6.1

```
In[39]:= (* generating the frames of a movie *)
    frames = Table[
        ArrayPlot[
         JuliaData[c, {{-2, 2}, {-2, 2}}, 200, 200]
        ],
        {c, 0, .75 I, .05 I}
        ];

In[44]:= (* displaying the frames in a grid *)
    k = 5; (* frames per row *)
    r = Mod[Length[frames], k];
    framesArray = Partition[frames, k];
    If[r > 0,
     AppendTo[framesArray, Take[frames, -r]];]
    GraphicsGrid[framesArray]
```

Out[47]=

of *list* and placing them in an array whose rows each have *k* elements. The problem with **Partition** is that if any elements of *list* are leftover they are simply dropped. In this case, we generated 16 frames, so if we place them in an array with rows that are five frames long, there is one frame leftover. We have written the example to take care of this, letting you choose *k* to be whatever you want. Note that **r**, defined in the second line, is the number of leftover frames. The third line makes the array, called **framesArray** using the **Partition** function and finally the fourth line tacks on a final partial line of frames if **r** is greater than zero.

To play the movie we use **ListAnimate** as seen in Example 10.6.2.

Example 10.6.2

In[230]:= (* using ListAnimate to play a list of
        frames *)
      ListAnimate[frames]

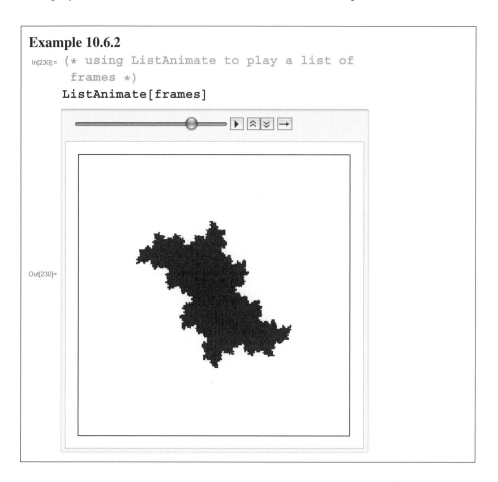

Out[230]=

ListAnimate will generate a **Manipulate** object containing an **Animator**. The controls can be used to stop or start the movie, play it faster or slower, and so on. All the same options available to **Manipulate** can be used with **ListAnimate**, with a few changes and additions. See the Help Files for more information.

If you are using the Mac OS X operating system it is also very easy to export the movie into a QuickTime movie. First, print all the frames of the movie using a **Do** loop as in Example 10.6.3. We have not included the output here because it will take up a lot of room. You need to do this yourself! Each frame will appear in its own cell, but the entire set of frames will have a single cell bracket enclosing them on the right. (Remember that, for most of the examples in this book, we have not been displaying the cell brackets.) Use the mouse to click on the cell bracket which surrounds all the frames of the movie, thereby selecting all of the frames. (Do *not* click on the largest possible cell bracket. This will select more than just the frames of the movie.) Now go to the **File** menu and select **Save Selection As...**. A dialog box will come up in which you choose QuickTime as the Format, name your movie and choose the location for the file as usual.[2]

Example 10.6.3

```
In[231]:= (* print the frames in separate cells in
    preparation for exporting the movie *)
Do[
Print[frames[[i]]],
{i, 1, Length[frames]}
]
```

10.7 The Mandelbrot Set

With Julia Sets we fixed the constant c and then looked at the orbits of *all* points. To generate the famous Mandelbrot Set, we'll vary c but only look at the orbit of *one* point, namely zero. The process is very similar to before. We'll take a region of the complex plane that we want to look at, subdivide it into little pieces, pick one point from each piece, use that point as c, and then follow the orbit of zero. If the orbit goes off to infinity the point is *not* in the Mandelbrot Set; if the orbit of zero does *not* go off to infinity, then the point *is* in the Mandelbrot Set. Example 10.7.1 defines the functions we need to plot the Mandelbrot Set. Let's look at the code and then discuss how it works.

The first function, **MandelbrotData** is similar to **JuliaData** except that we have added two features. We still enter the region of the plane that we want to look at, but instead of entering the number of divisions in both the horizontal and vertical directions, we now enter only one number of divisions (**div_**) which is used for the number of horizontal divisions (**xdiv=div**). The function then computes the number of vertical divisions with the instruction **ydiv=Floor[xdiv (ymax-ymin)/(xmax-xmin)]**. We use **Floor** here in case this quotient is not an integer. We have used a **Module** structure just so that we can do this, using **xdiv** and **ydiv** as local variables. By only

---

[2]We can also use this method to play the movie. After selecting all the frames of the movie, go to the **Graphics** menu, choose **Rendering** and then choose **Animate Selected Graphics**.

entering the number of horizontal divisions and letting *Mathematica* compute the number of vertical divisions we are now able to freely enter any size region and not have to worry about the plot coming out with the same scale in each direction.

The second feature of **MandelbrotData** is that we now output a list of two elements. The first element in the list is the region description {{**xmin, xmax**},{**ymin, ymax**}} and the second is the matrix of orbit lengths. Keeping the coordinates of the region and the data for that region together turns out to be handy. Aside from these two modifications, the guts of **MandelbrotData** are still very much like that of **JuliaData**, namely forming a **Table** of **orbitLength**.

Example 10.7.1

```
(*functions for drawing the Mandelbrot Set*)
MandelbrotData[
  {{xmin_, xmax_}, {ymin_, ymax_}}, div_] :=
 Module[{xdiv = div,
   ydiv =
     Floor[div (ymax - ymin) / (xmax - xmin)]},
  {{{xmin, xmax}, {ymin, ymax}},
   Table[orbitLength[0, x + I y],
     {y, ymax, ymin, -(ymax - ymin) / ydiv},
     {x, xmin, xmax, (xmax - xmin) / xdiv}]
  }
 ]
MandelbrotPlot[data_, colorfn_, xticks_,
  yticks_] :=
 ArrayPlot[data[[2]],
  ColorFunction → colorfn,
  DataRange → data[[1]],
  FrameTicks →
   {{Table[N[y], {y, ymin = data[[1, 2, 1]],
       ymax = data[[1, 2, 2]],
       (ymax - ymin) / yticks}], None},
    {Table[N[x], {x, xmin = data[[1, 1, 1]],
       xmax = data[[1, 1, 2]],
       (xmax - xmin) / xticks}], None}
   }
 ]
```

The second function, **MandelbrotPlot**, is basically **ArrayPlot** but with some options added. The first option is **ColorFunction**→*colorfn* where we have defined *colorfn* to be an argument of **MandelbrotPlot**. Thus the user needs to enter the desired coloring function when calling **MandelbrotPlot**. Any of the coloring functions that we defined in the last section could be used, and we give a few more suggested coloring schemes later. The second feature of **MandelbrotPlot** is to use the region parameters {{**xmin, xmax**},{**ymin, ymax**}} to place custom tick marks around the plot frame. This is done with the **FrameTicks**→{{*left, right*},{*bottom, top*}} option. Here we have provided lists of numbers for *left* and *bottom* (and **None** for *right* and *top*). This will place tick marks on the left and bottom edges of the plot at the indicated numbers. We use the **Table** function to create the lists, where the number of ticks on each edge are entered as arguments. Notice that the first time we get the value of *data[[1,2,1]]*, which is *ymin*, we name it **ymin** so that we can use the shorter name **ymin** the second time we need it, rather than having to type the longer *data[[1,2,1]]* again. This is only for convenience.

In the next example we plot the Mandelbrot Set with two different coloring schemes.

In the first cell of Example 10.7.2 we compute the Mandelbrot data (orbit lengths) for the region {{−2, 1}, {−1.1, 1.1}} using 1000 subdivisions (in the horizontal direction). In the second cell we define a custom coloring scheme that we call **grayScheme**. Finally, in the third cell we plot the data in two different ways. In the first plot we have used the coloring function **Automatic** so that we just use the default coloring scheme of **ArrayPlot**. In the second plot we use our own **grayScheme** color function.

The last cell also contains **Dynamic[MousePosition["Graphics"]]** which creates the very nice feature of printing the coordinates of the mouse whenever we move the mouse over a graphics object in the notebook. Try it out! As you move the mouse around inside the plot of the Mandelbrot Set, the position of the mouse, relative to the coordinates used to label the plot, are displayed where it initially says **None** in the output cell. If we had not used the **DataRange** option in the **ArrayPlot** function, then the coordinates would correspond to the position within the data array.

Using the mouse to read off coordinates in the plot is really useful for zooming in on the Mandelbrot Set. For example, it appears as though something interesting is going on near the point (−1.75, 0). Example 10.7.3 zooms in on this location.

Wow! A whole other miniature copy of the Mandelbrot Set! And could that little speck in this plot located at about (−1.786, 0) be another, even tinier copy of the Mandelbrot Set? Check it out!

Other interesting places to look are at (−.776, .125) or (−.162, 1.021). Virtually anyplace near the boundary of the solid black regions are worth zooming in on. You'll be surprised what you find! We won't do it here, but clearly it would be nice to make a movie of zooming in on some location. Just follow the examples of the last section, using **Table** to create the frames of the animation.

Example 10.7.2

```
In[51]:= (* drawing the Mandelbrot Set with two
            different coloring schemes *)
         data = MandelbrotData[
            {{-2, 1}, {-1.1, 1.1}}, 1000
            ];

In[52]:= grayScheme =
            If[# == 1.0, Black, GrayLevel[50 #]] &;

In[56]:= Dynamic[MousePosition["Graphics"]]
         MandelbrotPlot[data, Automatic, 2, 6]
         MandelbrotPlot[data, grayScheme, 2, 6]

Out[56]= None
```

Out[57]=

Out[58]=

Example 10.7.3

```
In[59]:= (* zooming in on the Mandelbrot Set *)
    size = .05;
    {x, y} = {-1.75, 0};
    data = MandelbrotData[
      {{x - size, x + size}, {y - size, y + size}},
      500
      ];
    MandelbrotPlot[data, grayScheme, 5, 5]
```

Out[61]=

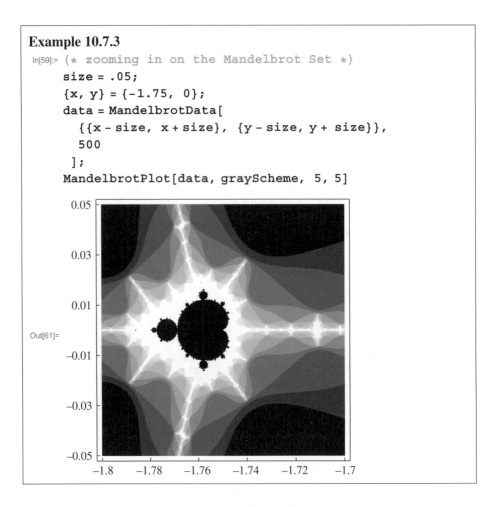

10.8 What Is a Fractal?

It is not easy to define what a fractal is and we will not attempt to give a precise definition. Generally speaking, *fractals* are sets that tend to have two important properties. The first is that they are *self-similar at all scales*. Basically what this means is that if we zoom in on a fractal it will continue to look essentially the same, no matter how far we zoom in. Our last example drives home this point about the Mandelbrot Set. After you spend some time exploring it you will discover that at every level of magnification there appear copies, or near copies, of the original set. We see the same phenomenon in the Julia Sets that we looked at. A certain spiral pattern might seem to always be present, no matter how far in we zoom.

The second hallmark of fractals is that they have *fractional dimension*. We are used to the concept of integral dimension: lines are one-dimensional, planes are two-dimensional, solids are three-dimensional, and so on. The number of dimensions of an object can be thought of as the number of "independent" directions it possesses, or equivalently, now many coordinates are needed to locate a point in it. How could something have a dimension of .562, or $\sqrt{2}$? But this can happen with Julia Sets! This seems really weird and indeed it is! Since this book is primarily about *Mathematica* and not about fractals, we are going to have to skip this rather fascinating topic of fractional dimension. We provide some references for further reading on fractals in the next section where you can learn more about these fascinating objects.

Even the Orbit Diagram that we encountered in the last section is a fractal! Notice that with each period doubling bifurcation, the set splits into two pieces, each of which bears a strong resemblance to the whole set. It is not exactly a miniature copy of the original but instead has the same basic shape.

10.9 Find Out More

We have barely scratched the surface of the theory of dynamical systems and fractals. There are literally hundreds of books on the topic. Three nice books to look at are

- *Chaos: Making a New Science* by James Gleick
- *The Fractal Geometry of Nature* by Benoit B. Mandelbrot
- *Chaos, Fractals, and Dynamics* by Robert L. Devaney

The first book gives a wonderful introduction to the history of dynamical systems and chaos theory. The second book is the groundbreaking work of Mandelbrot himself, and the third book contains nice mathematical investigations that you can pursue with *Mathematica*.

There are several nice demonstrations at the Wolfram site that explore Julia Sets and the Mandelbrot Set. A quick search of the Internet will undoubtedly turn up additional graphics and applications dealing with these topics.

In this section we discovered that speed of computation can sometimes be very important. We invite you to check out the tutorial "tutorial/CompilingMathematica Expressions" that can be found in the Help Files. And of course, looking up the references in the Help Files to the functions highlighted in this section will lead to options that we did not discuss as well as related functions.

Quiz

1. Use the following three examples to conclude that if p is a fixed point of $f(x)$ and $|f'(p)| = 1$, then p could be attracting, repelling, or neither. Use Graphical Analysis to decide.

   (a) Let $f(x) = -x$. Show that 0 is a fixed point that is neither attracting nor repelling. Show that $|f'(0)| = 1$.

   (b) Let $f(x) = \tan x$. Show that 0 is a repelling fixed point and that $|f'(0)| = 1$.

   (c) Let $f(x) = \sin x$. Show that 0 is an attracting fixed point and that $|f'(0)| = 1$.

2. Repeat Examples 10.3.6 or 10.3.7 to experimentally find the value of c where the attracting 2-cycle bifurcates to an attracting 4-cycle.

3. Let $f(x) = x^2 + c$. Use *Mathematica* to algebraically find the fixed points of $f(f(x))$ in terms of c. (Hint: **Solve** will work just fine.) Next, show that the slope of $f(f(x))$ at either of the period-2 points is $4 + 4c$. Conclude that the attracting period-2 cycle becomes repelling at $c = -1.25$.

4. Repeat Examples 10.3.6 or 10.3.7 to experimentally find the value of c where the attracting 4-cycle bifurcates to an attracting 8-cycle.

5. Plot the Julia set corresponding to $c = -.1 + .8i$. There are two ways to affect the resolution of the plot. The first is to vary the number of divisions, $xdiv$ and $ydiv$, in the horizontal and vertical direction. The second is to reset the maximum number of iterations in the **orbitLength** function. Try setting this as low as 50 or as high as 1000. What happens to your plots? Of course allowing more iterations as well as more horizontal and vertical divisions can greatly increase computation times.

6. Make a movie of the changing Julia Set as c moves around the circle of radius .8 centered at zero. (Hint: Let $c = .08e^{it}$ and then let t run from 0 to 2π in steps of $\pi/10$.)

7. Make a movie of the Mandelbrot Set as you zoom in on a particular point. To make the frames, include something like Example 10.7.3 in the body of a **Do** loop where you let the parameter **size** act as the counter for the loop.

8. In this entire chapter we have focused on the family of functions $f(z) = z^2 + c$. What if we switch to $f(z) = z^3 + c$, or perhaps to $f(z) = z^4 + c$? What if we change away from polynomials altogether and investigate $f(z) = \sin z$, or perhaps $f(z) = e^z$? Rewrite **orbitLength** in order to investigate other functions. However, can we still use the fact that an orbit that makes it outside the disk of radius 2 is headed for infinity?

CHAPTER 11

Looking Good

Until now we have focused on *doing* mathematics with *Mathematica*, either performing calculations or displaying results, textually or graphically. All of this has been out of context, although we have tried to give examples that relate to problems that might arise in your actual use of *Mathematica*. But writing mathematics is *much* more than just displaying a bunch of calculations one after the other. Good mathematical exposition reads much like a good story: there is a beginning, a middle, an end—even a punch line. Care must be taken to introduce all the right ideas in just the right order, with unnecessary details left out. And special attention must often be given to notation, which should be sufficient to describe the objects at hand, but not so complicated as to overwhelm the reader. In presenting our work, the right organization and selection of graphics can make the difference between striking out or hitting a home run.

In this chapter we'll see how a *Mathematica* notebook can contain all the elements needed for a self-contained mathematical exposition. Notebooks can be organized just as books are, with chapters, sections, and subsections. They can not only contain the calculations that we have been learning to make, but also text and other graphics. In addition, hyperlinks can be inserted anywhere in a notebook that can

lead elsewhere in the same notebook, or to other notebooks or even Web sites. Finally, notebooks can be used as slideshows for professional presentations.

11.1 Cell Styling

Mathematica provides a number of predefined "styles" that you can use to make your notebooks look certain ways. When a new notebook is opened the **Default Stylesheet** applies. But other styles such as **Article, Book**, or **Report** can be chosen from the **Format ▶ Stylesheets** menu item. When different styles are chosen, different options will then be available under the **Format ▶ Style** menu. For example, with the **Default** style, the **Format ▶ Style** menu lists 16 items, including **Title, Subtitle, Subsubtitle**, and so on. Try choosing **Format ▶ Stylesheets ▶ Book ▶ Textbook** and then look again at the **Format ▶ Style** menu. You'll now see a list of 40 items!

Each cell in a *Mathematica* notebook has a certain style and you may choose the style for a given cell from the **Format ▶ Style** menu. So far we have primarily discussed **Input** and **Output** cells. But to turn your notebook into a beautifully typeset expository article complete with mathematical calculations, you are going to want to use **Text** cells as well as an assortment of **Title** and **Subtitle** cells, and so on.

The style of any existing cell can be changed by selecting the cell (click on its bracket) and then choosing a style from the **Format ▶ Style** menu. The default is the **Input** style, used for entering mathematical expressions. Before creating a new **Text** cell, it is easiest to place the cursor where the cell will go, click the mouse, use the keyboard shortcuts to select the style, and then start typing. If you forget and start to type text into an **Input** cell, you can select the cell when you are done and change the format.

Suppose we want to create a notebook all about Pythagorean triples, sets of integers $\{x, y, z\}$ such that $x^2 + y^2 = z^2$. To create a nice title for the notebook, let's use a **Title** cell. The result is shown in the next example.

Example 11.1.1
```
(* a title cell *)
```
Pythagorean Triples

It is easy to change the title in a number of ways. In the next example we show the title after changing the font, the font size, the color, the background color, and the text alignment (which we changed to **Align Center**). All of these choices are easily found in the **Format** menu. Simply select the title cell and then select from the menu. Pretty cool title, eh?

Example 11.1.2

`(* a fancy title cell *)`

PYTHAGOREAN TRIPLES

After the title we will want an introductory paragraph. The following example shows four cells. The first is the **Title** cell, then a **Text** cell, then a **DisplayFormulaNumbered** cell, and finally, another **Text** cell. For the third cell, we could have just used a **DisplayFormula** cell, but as is often the case in mathematical exposition, we might want to number an equation so that we can refer back to it later.

Example 11.1.3

PYTHAGOREAN TRIPLES

One of the most famous theorems from antiquity is the Pythagorean Theorem. It is named for the Greek geometer Pythagoras, but was known to the Babylonians long before the Greeks, and also known independently to the Chinese. It states that in any right triangle there is a relationship between the lengths of the three sides. In particular, if the lengths of the two legs are a and b, and the length of the hypotenuse is c, then

$$a^2 + b^2 = c^2 \qquad\qquad (1)$$

For some right triangles, it is possible for all three sides to have integer lengths. That means that there are integers a, b, and c which satisfy Equation 1. Such a triple of integers is called a *Pythagorean Triple*. A triple known to most students, as well as all carpenters, is the 3, 4, 5 triple. (Carpenters use this to lay out right triangles so that walls meet at right angles.)

There is an important aspect to Example 11.1.3 that is not obvious. Notice that the mathematical symbols a, b, and c are not in the same font as the text.

Usually we mix text and formulas in a notebook by placing the different material in different cells. But it is possible to mix them within the same cell. To embed a mathematical expression within text, type Ctrl+(to begin the expression and then Ctrl+) to end it. (We can also end it by using the right arrow key.) When we are entering the mathematical expression we may use the BasicMathInput Palette or associated keyboard shortcuts. When the introductory paragraph in Example 11.1.3 was typed, the expressions a, b, and c were inserted this way, and when the formula $a^2 + b^2 = c^2$ was typed, the BasicMathInput Palette was used to set the exponents.

After entering the displayed formula, we centered the formula by selecting **Text Alignment ▶ Align Center**. We also selected the phrase "Pythagorean Triple" and reset it in italics. All the usual word processing tools are available.

By using cells with different styles, it is possible to make your notebook as polished as you like. If you are a college student using *Mathematica* to write your homework assignments, your professors will be thrilled! (And you may get a better grade!)[1]

11.2 The Option Inspector

We have already seen in Chap. 3 that we can use the **Option Inspector** to change a variety of preferences for the notebook. We can also use it to change all sorts of features related to the style of cells and the notebook as a whole.

Suppose, for example, we want to place a frame around a cell for added emphasis. Example 11.2.1 shows such an example.

Example 11.2.1

```
(* adding a frame around a cell *)
```
Section 1: Area

The area A of a disk of radius r is given by the famous formula $A = \pi r^2$.

```
(* calculation of area *)
r = 1.234;
A = π r^2
```
4.78388

---

[1]However, your professors will *not* be thrilled if you type text into input cells! Adding the explanatory text is great, but make sure it is placed in *text* cells.

One way to do this is to first select the cell and then choose **Format ▶ Option Inspector...** which will open up the **Option Inspector**. As mentioned in Chap. 3, the items in the **Option Inspector** can be displayed in different ways by choosing "by category" or "alphabetically" from the menu at the top of the **Option Inspector**. If we work alphabetically, simply scroll down until you see the option "CellFrame" which will currently be "False." Check the checkbox in the "Set" column (the first column) and then click on the word "False." This will allow you to edit the field and change "False" to "True." After you do this and hit the Return key, the frame will be inserted around the cell.

If you really want to look under the hood with *Mathematica*, you'll love the **Show Expression** command that can be found under **Cell** in the menu bar. In the following example we have placed a frame around a text cell that contains the text "Sample text." The very same cell is also shown after selecting it and then choosing **Cell ▶ Show Expression**. With this view we can see the underlying structure of the cell. Since this is an introductory book, we have not tried to explain this underlying structure in general. But notice that the cell contains the option **CellFrame→True**. An alternative way to place frames around cells is to first use **Show Expression**, then add the option **CellFrame→True**, and finally use **Show Expression** again to return to the normal view. You should try experimenting with **Show Expression**. Really complicated cells are going to look *really* complicated when you look at the underlying structure, but with a little practice you can start to see what is going on. If you find a cell in someone else's notebook that does just what you want, and you don't know how it works, using **Show Expression** might be just what you need to figure it out.

Example 11.2.2

```
In[1]:= (* using Show Expression *)
```

Sample text.

```
Cell["Sample text.", "Text",
CellFrame->True,
 CellChangeTimes->{{3.421059489087907*^9, 3.421059494393612*^9}}]
```

Returning to the **Option Inspector**, notice that you can choose from **Selection**, **Selected Notebook**, or **Global Preferences** from the pull-down menu at the top left. In the case of **CellFrame** for example, these three choices would put a frame around the selected cell, every cell in the current notebook, or every cell in all our notebooks, respectively. (However, some types of cells may have a particular style that would override the placement of a frame around the cell.)

You should experiment with the different built-in choices of *Stylesheets* that can be found under the **Format ▶ Stylesheet** menu item. For example, try **Format ▶**

Stylesheet ▶ Creative ▶ NaturalColor for a colorful presentation! Finally, it is possible to create your own Stylesheet for a completely custom style. This topic is a little beyond what the beginner would want to tackle, but digging into the Help Files will lead you to the necessary information.

11.3 Cell Properties

In addition to having different styles, cells may have different *properties*. Select a cell and then choose **Cell ▶ Cell Properties**. A list of five different properties, **Open, Editable, Evaluatable, Active**, and **Initialization Cell** will be shown, with check marks next to those that apply to the selected cell.

The first property is **Open**. We have already seen that by double-clicking a cell bracket you can "open" or "close" the cell, thereby showing or hiding its contents. Usually we would only want to close a cell group, rather than a single cell.

The next property is **Editable**. By unchecking this property we can make a cell uneditable, that is, "write-protected." This may be a good thing to do to cells that you want to make sure you don't accidentally change.

While you might want to change the first two properties of some cells you will probably never want to change whether a cell is **Evaluatable** or **Active**. Basically, you want all **Input** cells to be **Evaluatable** and all other cells to not be **Evaluatable**, and this is the default. So you never really need to mess with this property. Similarly, you may never need to change the **Active** status of a cell.

Typically you may work with a notebook for several hours or days as you develop it, often over more than one session. Once it is done, you may put it away for months before you need it again. When you close and reopen a notebook, you will have to reenter, or reevaluate, all the input cells. You could go through the notebook one cell at a time entering each one individually. Alternatively, you can select one or more cells and then choose **Evaluate ▶ Evaluate Cells** from the menu bar. Better still, as you write the notebook you can designate a cell as an *initialization cell*. To do this, select the cell and then select **Cell ▶ Cell Properties ▶ Initialization Cell** from the menu bar. After opening a notebook, choose **Evaluate ▶ Evaluate Initialization Cells**, and all the initialization cells will be automatically evaluated.

By now you may have noticed that cells with different styles and properties have different style cell brackets. You'll probably get to where you can recognize whether a cell bracket is open or closed based on the style of the bracket, and this is pretty useful. You may even learn to distinguish between a few other types of cells based on the appearance of the bracket. This can't hurt, but you certainly shouldn't feel that you need to learn all the different kinds of cell brackets.

11.4 Using Palettes

If you really want to make your notebook look good, then you will want to use the various palettes to typeset mathematical expressions in your input cells as well as set formulas within text. We have already talked about the BasicMathInput Palette as well as the AlgebraicManipulation Palette. In addition, you might find some characters you need on the SpecialCharacters Palette. And of course, the keyboard shortcuts available for many of the palette items are incredibly useful.

It can be annoying if you find yourself always using the same three or four palette items, but having to take them from different palettes. A nice solution is to create your own custom palette! You might even want to create your own custom palette item. In the next example we create a very simple palette that contains only one item. When this cell is evaluated, the *very* tiny palette will appear in the upper right corner of the screen. Try it!

Example 11.4.1

```
In[71]:= (* a simple custom Palette *)
    CreatePalette[
      PasteButton[ æ]
    ];
```

The function **CreatePalette** either takes a single expression as its argument, or a list of expressions, and creates a palette with these expressions as items. In this case we have only the single expression, **PasteButton**, which creates a button that will paste something into the notebook at the selection point when the button is pressed. In this case the special letter æ will be pasted into the notebook. If anything is selected when the button is pressed, the selection will be *replaced* by æ. Otherwise, æ will simply be *inserted* at the location of the cursor. By the way, when we typed the input cell of Example 11.4.1 we pulled the æ character off of the SpecialCharacters Palette!

Since we want some action to take place when we select from the palette, the argument to **CreatePalette** is typically either **Button** or **PasteButtton**. With the **Button** function we can carry out all sorts of operations when the button is pressed, not just pasting something into the notebook.

The next example is more complicated. We create a palette with four items, each a **PasteButton**. The first one will paste the special character $\mathbb{Z}$ into the notebook and is similar to Example 11.4.1. But the next three will paste in a template of a certain form, with blanks that we can fill in. The first two can be found on the ordinary MathInput Palette, but the last one is a new, custom palette item. Furthermore, the

Figure 11.1   A custom palette made with four **PasteButtons**.

four **PasteButtons** are the elements of a list that is the single argument of **Row**. This will cause the four buttons to be arranged nicely in a row. Finally, while **Row** is the single argument of **CreatePalette**, we follow it with the option **WindowTitle** → **"My Palette"** in order to place a title on the palette. When this cell is evaluated, the palette shown in Fig. 11.1 will appear on the screen. (Try replacing **Row** with **Column**. The buttons will be arranged vertically on the palette, but the title is now too wide to appear.)

Example 11.4.2

```
In[87]:= (* creating a custom Palette *)
    CreatePalette[
      Row[
        {
          PasteButton[ℤ],
          PasteButton[■□],
          PasteButton[■/□],
          PasteButton[
            DisplayForm[
              SubscriptBox[■, RowBox[{□, ",", □}]]
            ]
          ]
        }
      ],
      WindowTitle → "My Palette"
    ];
```

Creating the second, third, and fourth palette items in Example 11.4.2 was a bit tricky. To type the expression **PasteButton[■□]**, first type **PasteButton[**. Next, go to the BasicMathInput Palette and select **■□**. You might think that you could then finish typing the closing bracket and we would be in good shape. But think about it, by using the BasicMathInput Palette you simply inserted a *template* where you were typing. We still need to "fill in" the template! What we fill it in with are the special characters \[**SelectionPlaceholder**] and \[**Placeholder**]. So, after selecting the **■□** template, type "\[**SelectionPlaceholder**]." This will "fill in" the "base" of the template with "\[**SelectionPlaceholder**]." Next, hit the tab key, which will take you to the second placeholder in the template. Now type "\[**Placeholder**]." After this, use the right arrow to leave the template and then type the final square bracket. Interestingly, after you are done entering this, it does *not* look like

$$\textbf{PasteButton[\textbackslash[SelectionPlaceholder]}^{\textbackslash[Placeholder]}\textbf{]}$$

which *is* what you typed! (And you used the **■□** template to help you type it.) Instead, *Mathematica* turns the special characters into the filled and unfilled squares as soon as you complete typing them.

Once you understand how to create the second palette item, the third one should make sense too. The fourth one is slightly more complicated. We want to create the template **■□,□** so that we can easily enter something like $m_{1,2}$. We use **Subscript-Box** to create the basic form of the template, but need to wrap **DisplayForm** around it to make it look the way it should on the palette. We also use **RowBox** to lay out the two placeholders and the comma between them in the subscript. And again, we type \[**SelectionPlaceholder**] and \[**Placeholder**] to create the two different kinds of placeholders. We'll let you read about **DisplayForm, SubscriptBox**, and **RowBox** in the Help Files. These functions (as well as making custom palettes!) are taking us a little far from *Mathematica* for *beginners*.

By using the **Button** function in a palette, you can create palette items that do all sorts of cool things to whatever you may have selected in the notebook. Take a look at the **Applications** section of the **CreatePalette** Help File page where you will find code that creates a palette with two buttons labeled "Shout!" and "Quiet." Pressing the first button will change selected text to all uppercase and pressing the second will change it to all lowercase.

11.5 Cell Grouping

By now you have noticed that *Mathematica* automatically groups cells together in a hierarchical way. This hierarchy is based upon the styles of the cells. In Example 11.5.1 we have used two **Title** cells, four **Section** cells, five **Subsection** cells, and finally, two **Subsubsection** cells. Notice that everything following the first

Title cell (The Main Title) is grouped together, until we get to the second **Title** cell (Another Main Title). When the second **Title** cell is entered, *Mathematica* automatically starts a new cell group. Until that point, all cells are automatically placed inside the first cell group. The same is true at each lower level in the hierarchy.

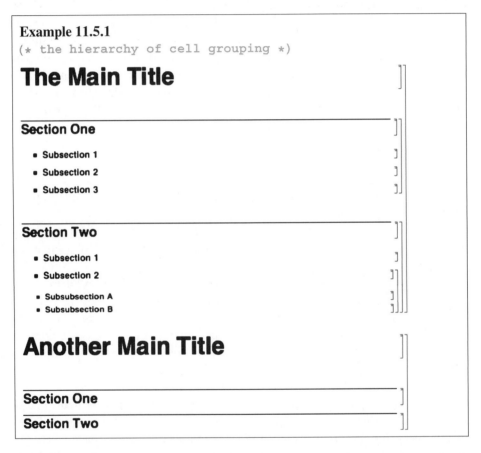

Example 11.5.1
(* the hierarchy of cell grouping *)

The Main Title

Section One

- Subsection 1
- Subsection 2
- Subsection 3

Section Two

- Subsection 1
- Subsection 2
  - Subsubsection A
  - Subsubsection B

Another Main Title

Section One
Section Two

A nice outcome of this automatic grouping is that we can close groups of cells so as to only present the headers of various sections. For example, closing the two highest level groups of cells results in Example 11.5.2.

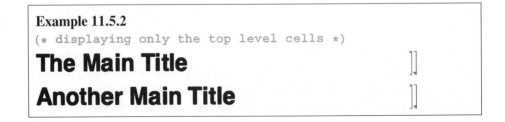

Example 11.5.2
(* displaying only the top level cells *)

The Main Title
Another Main Title

It is possible to exercise complete control on how cells are grouped by choosing **Cell ▶ Grouping ▶ Manual Grouping** (as opposed to **Cell ▶ Grouping ▶ Automatic Grouping**). With **Manual Grouping** chosen, no cells will be automatically grouped together. To group cells you must then select them and choose **Cell ▶ Grouping ▶ Group Cells/Group Together**. Cells can also be ungrouped by choosing **Cell ▶ Grouping ▶ Ungroup Cells/Group Normally**. In order for this to work, though, you need to have chosen **Cell ▶ Grouping ▶ Manual Grouping**.

It is also possible to merge two adjacent cells into a single cell, or to split a cell apart. To do the former, select the two cells and then choose **Cell ▶ Merge Cells**. To split a cell, place the cursor where you want the split to take place, then choose **Cell ▶ Divide Cell**.

Most of the time the Automatic Cell Grouping feature is just what you want. But if you need to change how your cells are grouped, you can take complete control.

Finally, we may wish to number our sections and subsections, and we can have *Mathematica* do this automatically. In Example 11.5.3 we have replaced the Title cells with Chapter 1 and Chapter 2, and the Section cells with Section 1, Section 2, and so on.

Example 11.5.3
```
(* using automatic cell numbering *)
```

Chapter 1

Section 1

Section 2

Chapter 2

Section 1

Section 2

The neat thing about this is that we did *not* type "Chapter 1" into the first Title cell! Instead, we typed "Chapter " and then chose **Insert ▶ Automatic Numbering....** This brought up the dialog box seen in Fig. 11.2, where we then chose "Title" for the **Counter**. The *really* cool feature of choosing **Automatic Numbering** is that if

Figure 11.2 Choosing **Insert ▸ Automatic Numbering...** brings up this dialog box. Using "Title" as the counter will number the insertion according to the current title number.

we were to rearrange the order of the Chapters (or Sections, Subsections, etc.) by cutting and pasting, *Mathematica* would automatically renumber them!

11.6 Hyperlinks

It is possible to insert hyperlinks into a *Mathematica* notebook that can be used to jump to a cell anywhere in the same notebook, another *Mathematica* notebook, or even any webpage. For example, suppose in our *Pythagorean Triples* notebook we want to have a link to the Wikipedia webpage on the subject. The URL of the webpage is

http : //en.wikipedia.org/wiki/Pythagorean_theorem.

To create the link, select the text that will serve as the hyperlink and then choose **Insert ▸ Hyperlink...** from the menubar. In Example 11.6.1, we have chosen

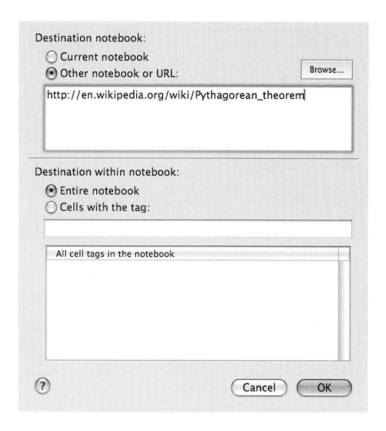

Figure 11.3   To add a hyperlink, select the text to serve as the link and then choose "Hyperlink…" from the "Insert" menu. Then fill in the dialog window accordingly.

"Wikipedia article." This will bring up the dialog box shown in Fig. 11.3. Click the "Other notebook or URL:" button and then enter the URL in the field as shown in the figure. The text cell with the hyperlink will then appear as in Example 11.6.1. Note that the hyperlink is given a different color from the ordinary text.

Example 11.6.1

`(* creating a hyperlink *)`

For more information on the Pythagorean Theorem, we recommend the Wikipedia article.

If instead of jumping to a webpage, we want to jump to another *Mathematica* notebook, choose "Current notebook" in the dialog box and then enter the name of the notebook in the field. The easiest way to do this is to use the "Browse..." feature, which will allow you to choose the notebook from your computer's file system. When you do this you will notice that the complete path name to the file is used, either in absolute or relative terms. If this last sentence doesn't make sense, don't worry. Just use the "Browse..." feature to select the correct notebook. *Mathematica* will insert the correct file name in the field. Clicking the hyperlink will then open the destination notebook.

Alternatively, we may want a hyperlink that takes us to a specific cell elsewhere in the same notebook. To do this we need to give a *tag* to the cell we wish to jump to. For example, suppose we wanted to include the following sentence in our "Pythagorean Triples" notebook.

Example 11.6.2
```
(* creating a hyperlink *)
```
A proof of the Pythagorean Theorem is given later in this notebook.

In Example 11.6.2 the text "proof of the Pythagorean Theorem" will be the hyperlink that takes us to the first cell of Example 11.6.3. We first select the initial cell in Example 11.6.3, and then choose **Cell ▶ Cell Tags ▶ Add/Remove Cell Tags...**. This will bring up a dialog box used for editing cell tags. We enter the name we want to assign to the cell as its tag into the **Cell Tag:** field and click **Add**. In this example, we used the name "proofOfPythagoreanThm" for the tag. We can now create a hyperlink anywhere else in the notebook that will point to this cell. As before, select the text that will serve as the link and then choose **Insert ▶ Hyperlink...** to bring up the hyperlink dialog box. This time choose **Cells with the tag:** and then either type the tag name into the field, or select the tag name from the list of all tag names. Then click **OK**. Clicking on the hyperlink will now take us to the tagged cell.

In Example 11.6.3 the tag name appears just above the tagged cell. This is because we have selected **Cell ▶ Cell Tags ▶ Show Cell Tags**. This feature can be turned on or off at any time.

If you are working with a really large notebook, you may want to use hyperlinks to create a "table of contents" at the beginning of the notebook. Each entry in the table can be a hyperlink to the corresponding section later in the notebook.

Example 11.6.3
```
(* giving a tag to a cell *)
```

proofOfPythagoreanThm

Proof of the Pythagorean Theorem

To prove the Pythagorean Theorem, suppose that T is a right triangle with legs of lengths *a* and *b* and hypotenuse of length *c*. Four copies of T may be arranged as shown below, forming a large square of side *c* with a square hole in the middle. The side of the square hole has length *b* − *a*. If we now express the area of the big square in two ways, either at c^2, or as the sum of the areas of the four triangle plus the area of the square hole, we obtain

$$c^2 = 4\left(\frac{ab}{2}\right) + (b-a)^2 = 2ab + b^2 - 2ab + a^2 = a^2 + b^2.$$

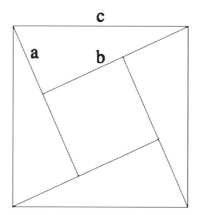

11.7 Adding Graphics to Notebooks

In Chaps. 2 and 6 we saw how to use *Mathematica* to create all sorts of graphical output. Obviously we could use these graphics to illustrate our notebooks. For example, the figure in Example 11.6.3 was created with *Mathematica*. But notice that we do not display the input cell that created the graphic. We have also discarded the output cell label "Out[n]=" that originally accompanied the output. To do this we can create the graphic in another notebook, save only the graphic and then paste the graphic into the notebook where it will be used. Let's see how to do this.

First, in another notebook, we create the graphic with the following input cell.

Example 11.7.1

```
In[21]:= (* graphic for proof of Pythagorean Theorem *)
    (* (x, y) are coordinates of right-
    most vertex of inner square *)
    y = -1 / 4; x = Sqrt[1 - (y + 1) ^2];
    Show[Graphics[
      {(* draw outer square *)
      Line[{{-1, -1}, {1, -1}, {1, 1}, {-1, 1},
        {-1, -1}}],
      (* draw four diagonal lines *)
      Line[{{-1, -1}, {x, y}}],
      Line[{{1, -1}, {-y, x}}],
      Line[{{1, 1}, {-x, -y}}],
      Line[{{-1, 1}, {y, -x}}],
      (* place text labels *)
      Style[Text["c", {0, 1.1}], 24],
      Style[Text["b", {0, .65}], 24],
      Style[Text["a", {-.75, .7}], 24]
      }
    ]
  ]
```

Out[22]=

Next, click on the graphic to select it and then choose **Edit ▶ Cut**, or use the keyboard equivalent, to copy the graphic. We can then **Paste** it into any other notebook at the desired location. If you select the cell bracket rather than the graphic, cutting and pasting will include the "Out[n]=" label, which you may not want to include.

Alternatively, after selecting the graphic we can save it as a file by selecting **Edit ▸ Save Selection As....** We'll see in a moment how to import a graphics file into a notebook. A third option is to select the graphic and then choose **Edit ▸ Copy As ▸ PDF** or **Edit ▸ Copy As ▸ PICT** to again save the graphic as a file in either PDF or PICT format. You'll need to experiment with the different methods and different graphics formats to see what works best for any specific graphic.

Instead of creating your own graphic with *Mathematica*, you may want to paste a graphic obtained from some other source into a *Mathematica* notebook. In the following example we first used a screenshot utility to capture the *Mathematica* notebook logo and then saved it as a file named "notebookLogo.png." We then pasted the graphic into the notebook by selecting **Insert ▸ Picture ▸ From File....** Selecting this from the menu bar brings up a file directory from which you can choose any file. After choosing the desired file and clicking **Open**, the picture is pasted into the notebook.

Example 11.7.2

```
(* selecting Insert→Picture→From File...
   to insert a picture *)
```

We can accomplish the same thing, but with more control, by using the **Import** function. The basic syntax for the **Import** function is **Import[***"filename.ext"***]**. Notice that in Example 11.7.3 we have given the complete path name to the file. We have also used the optional **ImageSize** to control the size of the image.

Example 11.7.3

```
In[30]:= (* importing a graphics file *)
    Import[
     "/Users/jhoste/Desktop/MathDemystified/Chapters
       /LookingGood/notebookLogo.png",
     ImageSize → {50, 50}]
```

Out[30]=

In the next example we import a digital photograph saved in the .jpg format and exert a little more control over its size. The first use of **Import** uses the optional

Mathematica Demystified

"ImageSize" element and returns the size of the image, in this case $\{2816, 2112\}$. We name the size **photoSize**. Next, we use **Import** again, this time with the optional **ImageSize** which we set to 15% the size of the photo.

Example 11.7.4

```
In[3]:= (* finding the size of an imported image *)
    photoSize =
     Import[
       "/Users/jhoste/Desktop/MathDemystified/
         Chapters/LookingGood/medusa.jpg",
       "ImageSize"]

Out[3]= {2816, 2112}

In[4]:= (* importing the image and scaling its size *)
    Import[
      "/Users/jhoste/Desktop/MathDemystified/Chapters
        /LookingGood/medusa.jpg",
      ImageSize → .15 photoSize]
```

Out[4]=

In addition to importing graphics, **Import** can be used to import all sorts of data. Take a look at the Help Files under **Import**.

11.8 Creating Slideshows

It is easy to create slideshows using *Mathematica*. There are a couple of different ways to get started, but a nice way is to use the SlideShow Palette seen in Fig. 11.4. First open a new notebook and then open the Palette by choosing **Palettes ▶ SlideShow** from the menu bar. Alternatively, instead of opening a new notebook, you could choose **File ▶ New ▶ SlideShow** from the menu bar, rather than **File ▶ New ▶ Notebook**. This will open a notebook in the form of a slideshow with three already existing slide *templates* that you can then edit. Try it out! This may end up being your preferred method, but by starting with a new notebook and using the Palette you won't need to replace the three existing slides. (Yet another method is to first open the Palette and then click **New Template**. This will produce the same result as **File ▶ New ▶ SlideShow**.)

Assuming we have started with a brand new notebook and then opened the SlideShow Palette, click **New Slide**. This will insert a single slide template as seen in Fig. 11.5. This slide template contains three cells which are grouped as seen in the figure. The first cell is a **Navigation Bar** and the last cell is a **Previous/Next**

Figure 11.4   The SlideShow Palette can be used to create slideshows.

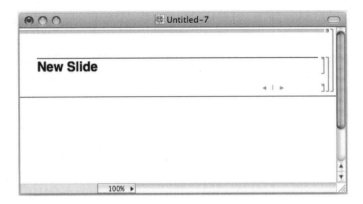

Figure 11.5 Clicking **New Slide** in the SlideShow Palette will paste a slide template into the notebook.

cell. These can actually be pasted into the notebook individually by selecting from the **Paste...** menu in the **Extras...** section of the Palette. The middle cell is simply a **Section** cell, and of course, it also could have been pasted in individually.

With the first slide template in place, we can then edit the **Section** cell and add whatever other cells we want. After finishing the content of the first slide, we can then click **New Slide** again to produce the second slide and so on. In Fig. 11.6 we see the first two slides of a slideshow about Pythagorean Triples. A combination of **Text** cells and **DisplayFormulaNumbered** cells have been used. In some of the **Text** cells we have also inserted typeset mathematical content as described in Sec. 11.1.

After preparing the slides we'll want to switch the **Screen Environment** to **SlideShow** to view the slides, or to use them in our slide presentation. There are two ways to do this. The first way is to choose **Format ▶ Screen Environment ▶ SlideShow**. The second way is to click **SlideShow** on the SlideShow Palette. Clicking **SlideShow** and **Normal** on the Palette will take you back and forth between the **SlideShow** and **Working** Screen Environments. Figure 11.7 shows the third slide of the slideshow with the **Screen Environment** set to **SlideShow**. The buttons at the top of the slide can now be used to go forward or backward one slide at a time, or to jump all the way to the first or last slide. We can also do this by clicking the left or right buttons at the bottom of the slide. (These are actually hyperlinks.) We can also expand the slide to fill the screen by clicking the "full screen" icon at the top left.

The figure on the third slide was made "on the fly" by first choosing **Graphics ▶ New Graphic** to create a "blank picture." We then selected **Graphics ▶ Drawing Tools** to bring up the 2D Drawing Palette. The line tool was used to draw the lines and the oval/circle tool was used to draw the circle. (Note that holding down the

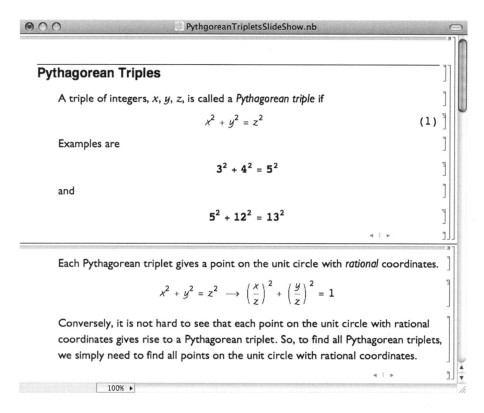

Figure 11.6   The first two slides of a slideshow about Pythagorean Triples. The notebook is seen in **ScreenEnviornment ▶ Working** format.

Shift key while using these tools will allow only horizontal or vertical lines, and circles, respectively.) The circle originally appeared as a filled-in disk and the 2D Graphics Inspector was then used to change the opacity of the interior of the disk. Finally the "TraditionalForm Text" as opposed to the "Text" tool was used to place the two labels. If the "Text" tool is used, the letters will not appear when the **Screen Environment** is changed from **Working** to **SlideShow**.

If we click **Table of Contents** on the SlideShow Palette, the table of contents shown in Fig. 11.8 will be generated. When giving the presentation it can be handy to have the Table of Contents visible on the screen. Not only will the audience have a sense of where you are in your presentation, it can be used to jump forward or backward to any slide in the presentation. At the end of the presentation when an audience member asks you to go back to an earlier slide, this avoids having to go back through them all one at a time.

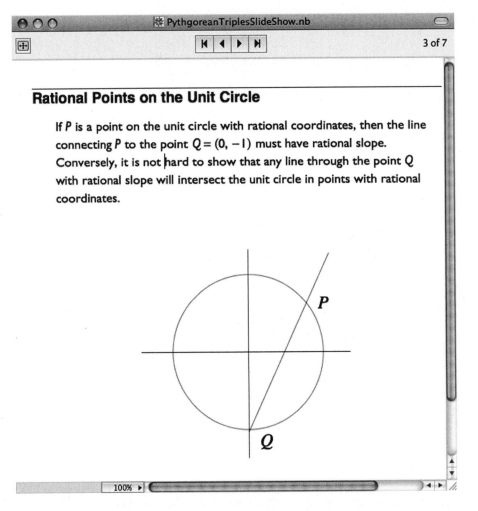

Rational Points on the Unit Circle

If P is a point on the unit circle with rational coordinates, then the line connecting P to the point $Q = (0, -1)$ must have rational slope. Conversely, it is not hard to show that any line through the point Q with rational slope will intersect the unit circle in points with rational coordinates.

100%

Figure 11.7 The third slide of a slideshow about Pythagorean Triples. The notebook is seen in **ScreenEnvironment ▶ SlideShow** format. Clicking the "full screen" icon at the top left will expand the window to the full computer screen. The navigation buttons at the top can be used to go forward or backward one slide at a time, or all the way to the first or last slide.

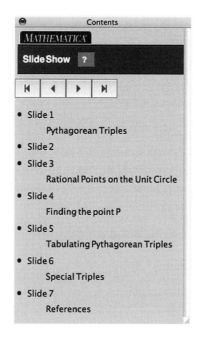

Figure 11.8   The Table of Contents for the slideshow. Clicking "Slide n" will move to that slide.

11.9 Find Out More

By now you have learned that *Mathematica* is almost infinitely customizable. To change the look and feel of your notebooks, the following guides and tutorials will take you beyond the basics. As usual, enter these phrases in the Documentation Center search field to bring up the desired guide or tutorial.

- guide/NotebookBasics
- guide/CellStylingOptions
- tutorial/OptionsForCells
- guide/AutomaticTextStylingFeatures
- tutorial/StylesAndTheInheritanceOfOptionSettings
- guide/Stylesheets

Quiz

1. Typeset the text cell shown in the following example. Make sure that the variables and the equation are all set as in-line formulas so that they appear correctly.

   Fermat's Last Theorem states that if x, y, and z are positive integers such that $x^n + y^n = z^n$ for some positive integer n, then n, must be 1 or 2.

2. Use the Option Inspector to decorate a text cell with the *filled star dingbat* as shown in the following example.

   ★ An integer is divisible by 3 if and only if the sum of its digits is divisible by 3.

3. Use the Option Inspector to change a cell background color to blue.
4. Use the Option Inspector to place a red cell frame around a single cell.
5. Use the Option Inspector to place red cell frames around *all* the cells in a notebook.
6. Create a custom palette with a single **PasteButton** that allows you to paste the special character "∞" into your notebook.
7. Create the custom palette shown below that allows you to paste in a 3×3 matrix.

8. Create the following text cell with a hyperlink to the United States Track and Field Web site at `http://www.usatf.org/`.

   A comprehensive listing of track and field records can be found at the United States Track and Field Web site.

9. Use **Import** to paste a digital photo into a notebook.
10. Make a SlideShow with three slides.

Final Exam

1. Name at least four types of delimiters used by *Mathematica* and describe how they are used.
2. Name two ways to hide large output cells.
3. What is the difference between $\sqrt{3}$ and $\sqrt{3.0}$?
4. What does **N[x]** do?
5. What does **?Select** do?
6. What does **?*String** do?
7. What is /. used for?
8. Explain the difference between := and =.
9. How is the ampersand, &, used?
10. How is the "at" symbol, @, used?
11. What happens if you take the square root of a *list* of 10 numbers?
12. What is the difference between {1, 2, 3}*{a, b, c} and {1, 2, 3}.{a, b, c}?
13. Plot the graphs of $y = x^2$ and $y = 10 - x^2$ simultaneously.
14. Shade the area trapped between the two curves in the previous question.

15. Use **Manipulate** to do the following: Plot the graph of $y = x^3 - 5x + 1$ and a point P somewhere in the plane that the user can move with the mouse. As the point is dragged about, plot the lines that pass through P which are tangent to the curve.

16. Plot the graph of the parabola $x = y^2$, whose axis of symmetry is the *horizontal* axis.

17. Plot the surface given by $z = x^2 - y^3$ over all points in the xy-plane located within 1 unit of the origin.

18. Plot the surface of revolution obtained by spinning the curve $y = \sqrt{x - 1}$ around the y-axis.

19. Plot the contours of the function $f(x) = \sin(x^2 - y^3)$.

20. Use **ContourPlot3D** to graph the surface

$$\left(z - \sqrt{x^2 + y^2 + 1}\right)(x^2 + y^2 + z^2 - 9) = 3$$

21. Use **Graphics3D** to draw a tetrahedron. Make each edge of the tetrahedron 10 units long. Place a sphere of radius 1 at each vertex, and place a cylinder of radius 1/2 along each edge.

22. Find the shape of the right circular cone that has the greatest volume for a given amount of surface area.

23. Find the center of mass of the Great Pyramid at Cheops. (Assume the pyramid is of uniform density.)

24. A pizza is removed from an oven at 450 degrees Fahrenheit and left to cool in the kitchen where the air temperature is 72 degrees. After 5 minutes the pizza has cooled to 300 degrees. Assuming that it cools at a rate proportional to the difference between its temperature and that of the air, when will it reach 200 degrees?

25. Make a list of the first 1000 prime numbers. Use the **Select** function to extract all primes from the list that are congruent to 1 mod 4.

26. *Twin primes* are primes that differ by 2. For example, 3 and 5 are twin primes as are 29 and 31. Find the first 100 pairs of twin primes. (It is a famous unproven conjecture that there are infinitely many pairs of twin primes.)

27. A number is *perfect* if it is the sum of its proper factors. Six is perfect since $6 = 1 + 2 + 3$, but eight is not perfect because $8 \neq 1 + 2 + 4$. Find the first four perfect numbers. (Hint: They are all less than 10000.)

28. Use a **Do** loop to print the first 100 Fibonacci numbers, each on a separate line.

29. Use a **While** loop to keep printing out successive powers of 2 (starting at $2^1 = 2$) until you reach a number bigger than 1000000.

30. Create an animation of a bouncing ball. Assume the ball falls with constant acceleration due to gravity from a height of 100 ft. Ignore air resistance. Ignore having the ball deform on impact, and assume that no energy is lost in the bounce, so that it bounces back as high as it fell. Simplify things by making the view from the side so that the ball appears as a disk bouncing on a horizontal line.

31. A *palindrome* is a word that reads the same forwards as backwards. For example, "radar" or "toot." Find all palindromes in the *Mathematica* provided dictionary.

32. We often hear that the surface of the earth is 70% water. Use the data in **CountryData** to estimate the fraction of the earth's surface that is covered by water.

33. Use **CountryData** to draw a map of Africa. Use **Tooltip** so that the name of each country pops up as the mouse is moved over the country.

34. Use **CountryData** to plot the GDP of the United States over the last 20 years.

35. Use **Fit** to fit a curve to the data of the last example. Use the fitted curve to predict the GDP 5 years from now.

Answers

Quiz 1

1.1

```
(1 / 2 + 1 / 3) ^ 3
```

$$\frac{125}{216}$$

1.2

```
(1 / 2 + 1 / 3) ^ 3.0
```

0.578704

1.3

```
N[E ^ Pi]
N[Pi ^ E]
```

23.1407

22.4592

1.4
```
(* radius of earth in miles *)
r = 4000
(* volume of earth in cubic miles *)
V = N[4 / 3 Pi r^3]
(* surface area in square miles *)
S = N[4 Pi r^2]
```

4000

2.68083×10^{11}

2.01062×10^{8}

1.5
```
(* volume of pyramid of Cheops *)
(* base area in square meters *)
baseArea = 230^2
(* height in meters *)
height = 147
(* volume in cubic meters *)
volume = 1 / 3 baseArea height
```

52 900

147

2 592 100

1.6
```
Table[k^3, {k, 1, 10}]
```
{1, 8, 27, 64, 125, 216, 343, 512, 729, 1000}

1.7
```
Table[N[Sin[x], 10], {x, 0, Pi / 2, Pi / 20}]
```
{0, 0.1564344650, 0.3090169944,
  0.4539904997, 0.5877852523,
  0.7071067812, 0.8090169944, 0.8910065242,
  0.9510565163, 0.9876883406, 1.0000000000}

Alternatively, **Sin** is *listable* so that we can take **Sin** of a list and it will apply **Sin** to each element of the list. Notice also using 0.0 to force approximations.

```
Sin[Table[x, {x, 0.0, Pi / 2, Pi / 20}]]
{0., 0.156434, 0.309017,
  0.45399, 0.587785, 0.707107, 0.809017,
  0.891007, 0.951057, 0.987688, 1.}
```

1.8

```
Table[N[(1 + 1 / 10 ^ k) ^ (10 ^ k)], {k, 1, 6}]
N[E]
{2.59374, 2.70481, 2.71692,
  2.71815, 2.71827, 2.71828}

2.71828
```

1.9

```
Table[n!, {n, 0, 20}]
{1, 1, 2, 6, 24, 120, 720,
  5040, 40320, 362880, 3628800,
  39916800, 479001600, 6227020800,
  87178291200, 1307674368000,
  20922789888000, 355687428096000,
  6402373705728000, 121645100408832000,
  2432902008176640000}
```

Quiz 2

2.1

```
Plot[-x^2 + x + 1, {x, -2, 2}]
```

2.2

```
Plot[-x^2 + x + 1, {x, -2, 2},
  AspectRatio → Automatic]
```

2.3

```
Plot[{-x^2 + x + 1, x / 2 - 1}, {x, -2, 2}]
```

2.4

```
Plot[{-x^2 + x + 1, x / 2 - 1}, {x, -2, 2},
  PlotStyle -> {Thickness[.01],
    Thickness[.03]}]
```

2.6

```
PolarPlot[1 + Cos[θ], {θ, 0, 2 π}]
```

2.7

```
Manipulate[
 PolarPlot[a + Cos[θ], {θ, 0, 2 π},
  PlotRange → {{-2, 3}, {-3, 3}},
  AspectRatio → Automatic],
 {a, -2, 2}
]
```

2.8

```
Graphics[
 Table[
  Rectangle[{i, i}, {i + 1, i + 1}],
  {i, 0, 5}
 ]
]
```

2.9

```
Graphics[
 Table[
  {GrayLevel[i / 10],
   Rectangle[{i, i}, {i + 1, i + 1}]},
  {i, 0, 5}
 ]
]
```

2.10

```
ListPlot[
 Table[
  RandomReal[1, 2],
  {1000}
 ],
 AspectRatio → Automatic
]
```

Quiz 3

3.1

```
Plot[
 E^x,
 {x, -1, 10},
 PlotRange → {{-1, 4}, {0, 10}},
 PlotLabel → "The Graph of e^x."
]
```

3.3

```
(*Sieve of Eratosthenes*)
(*initialize the
 sieve:make a list of the integers
   from 1 to max*)
max = 100;
sieve = Table[i, {i, 1, max}];
(*now for each integer k from 2 to
 max/2, see if it has been crossed
 out. If not,
cross out all of its multiples. A
 number in the list will be ''crossed
 out'' if it has been replaced with a
 zero *)
Do[(*see if the number k has been
   crossed out*)
  If[sieve[[k]] ≠ 0,
    (*cross out all its multiples*)
    j = 2;
    While[
      j k ≤ max,
      sieve[[j k]] = 0;
      j++
    ]
  ],
  {k, 2, max / 2}
]
(* the sieve now contains the primes
 and zeroes. Union will remove
 duplicate zeroes and sort the
 list. Drop removes the 0 and 1 that
 are now the first two elements*)
primes = Drop[
  Union[sieve]
  , 2
  ]
```

3.5

```
Monitor[
 Table[
   Pause[1],
   {i, 1, 10}
   ];,
  {i, FactorInteger[i]}
 ]
```

The integers and their prime factorizations are printed at 1 second intervals. Removing the semicolon causes the table to be printed after the calculation is complete. But the table does not contain anything!

Quiz 4

4.1

$$\text{PowerExpand}\left[\sqrt{\frac{w^3\, x^{-2}\, y^5}{w^5\, xz^3}}\,\right]$$

$$\frac{y^{5/2}}{w\, x\, xz^{3/2}}$$

4.2

```
Simplify[Sum[i^3, {i, 1, n}]]
```

$$\frac{1}{4}\, n^2\, (1+n)^2$$

4.3

```
TrigExpand[Sin[4 θ]]
```

$$4\, Cos[θ]^3\, Sin[θ] - 4\, Cos[θ]\, Sin[θ]^3$$

4.4

```
TrigReduce[-4 Sin[x]^3 + 4 Cos[x] Sin[x]^2 +
   3 Sin[x] - Cos[x]]
```

$$-Cos[3\, x] + Sin[3\, x]$$

4.5

```
Table[
  Expand[x^5 - 3 x^2 + 6 /. x → (x + k)], {k, -4, 4}
 ]
```

4.6

```
Table[
  Coefficient[(x² + x + 1)ⁿ, x, 3],
  {n, 2, 25}
]
```

4.7

```
Log[x y z] //. Log[a_b_] → Log[a] + Log[b]
```

$\text{Log}[x] + \text{Log}[y] + \text{Log}[z]$

```
Log[x / y] /. Log[a_ / b_] → Log[a] - Log[b]
```

$\text{Log}[x] - \text{Log}[y]$

Combining the rules gives

```
Log[x y z / w] //.
  {Log[x_ / y_] → Log[x] - Log[y],
   Log[x_ y_] → Log[x] + Log[y]
  }
```

$-\text{Log}[w] + \text{Log}[x] + \text{Log}[y] + \text{Log}[z]$

But note that changing the order does not work right!

```
Log[x y z / w] //.
  {Log[x_ y_] → Log[x] + Log[y],
   Log[x_ / y_] → Log[x] - Log[y]
  }
```

$\text{Log}\left[\dfrac{1}{w}\right] + \text{Log}[x] + \text{Log}[y] + \text{Log}[z]$

This will work:

```
Log[x y z / w] //.
  {Log[x_ y_] → Log[x] + Log[y],
   Log[x_ / y_] → Log[x] - Log[y],
   Log[1 / b_] → Log[1] - Log[b]
  }
```

$-\text{Log}[w] + \text{Log}[x] + \text{Log}[y] + \text{Log}[z]$

4.8

```
ReplaceRepeated[x, a_ → 1 + 1 / a,
  MaxIterations → 4]
```

4.9

```
data = Table[RandomReal[], {20}]
temp = Drop[
  data, Position[data, Min[data]][[1]]
  ]
Drop[
  temp, Position[temp, Max[temp]][[1]]
  ]
```

4.10

```
data = Table[RandomInteger[100], {10}]
data /. {data[[4]] → data[[5]],
  data[[5]] → data[[4]]}
```

4.11

```
N[Pi / 2]
```

1.5708

```
(* write product using two factors at
  a time *)
```

$$N\left[\text{Product}\left[\frac{(2\,k)^2}{(2\,k-1)\,(2\,k+1)}, \{k, 1, \#\}\right]\right] \& /@$$
$$\{5, 50, 500\}$$

{1.50109, 1.56304, 1.57001}

$$\text{Product}\left[\frac{(2\,k)^2}{(2\,k-1)\,(2\,k+1)}, \{k, 1, \text{Infinity}\}\right]$$

$$\frac{\pi}{2}$$

Quiz 5

5.1

```
surfaceArea[r_, h_] := 2 π r² + 2 π r h
```

5.2

$$\text{payment}[P\_, r\_, m\_] := \frac{P\,\frac{r}{12}\,\left(1+\frac{r}{12}\right)^m}{\left(1+\frac{r}{12}\right)^m - 1}$$

5.3
```
(* returns True if leap year,
False otherwise *)
leap[yr_] := If[Mod[yr, 400] == 0,
  True,
  If[Mod[yr, 100] == 0,
   False,
   If[Mod[yr, 4] == 0,
    True,
    False
   ]
  ]
 ]
```

5.4
```
(* days per months in non-leap years *)
daysInMonths = {31, 28, 31, 30, 31, 30,
   31, 31, 30, 31, 30, 31};

(* converts a date to the day in the year *)
date2day[{day_, month_, year_}] :=
 Total[Take[daysInMonths, month - 1]] + day +
  If[month > 2 && leap[year], 1, 0]
```

5.5
```
Map[#[[1]] Sin[#[[2]]] &, data]
```

5.6
```
n = 8;
title = "Multiplication table for 1 to " <>
   ToString[n];
Grid[{{title}, {
   TableForm[
    Table[i * j, {i, 1, n}, {j, 1, n}],
    TableHeadings → {
      Table[i, {i, 1, n}],
      Table[j, {j, 1, n}]
     }
    ]
   }}
 ]
```

5.7

```
(* distance from n to nearest prime *)
distanceToNearestPrime[n_] := Module[
   {bigger = n, smaller = n},
   While[PrimeQ[bigger] == False, bigger++];
   While[PrimeQ[smaller] == False, smaller--];
   Min[{Abs[smaller - n], Abs[bigger - n]}]
   ]
```

5.8

```
primesInInterval[n_, m_] := Module[
   {k = n, primes = {}},
   While[k <= m,
    If[PrimeQ[k], AppendTo[primes, k]];
    k++
    ];
   primes
   ]
```

5.9

```
n = 1;
While[collatzStoppingTime[n] ≠ 200, n++]
n
```

5.10

```
babylon[a_, x0_] := Module[
   {x = N[x0]},
   While[Abs[x^2 - a] > 1 / 10^5,
    Print[x]; (* optional print *)
    x = (x + a / x) / 2];
   x
   ]
```

Quiz 6

6.1

```
Plot3D[Cos[x] + Sin[y], {x, 0, 4 π}, {y, 0, 4 π}]
```

6.2

```
Plot3D[Cos[x] + Sin[y],
 {x, -π, π}, {y, -π/2, 3π/2},
 RegionFunction → Function[{x, y, z}, z ≥ 0]]
```

6.3

```
Plot3D[Cos[x] + Sin[y],
 {x, -π, π}, {y, -π/2, 3π/2},
 RegionFunction → Function[{x, y, z}, z ≥ 0],
 BoxRatios → {2π, 2π, 2}
]
```

6.4

```
Plot3D[Cos[x] + Sin[y],
 {x, -π, π}, {y, -π/2, 3π/2},
 RegionFunction →
  Function[{x, y, z}, z ≥ 0 && (x < 0 || y > 3π/4)],
 BoxRatios → {2π, 2π, 2}
]
```

6.5

```
ContourPlot[
 Cos[x] + Sin[y], {x, 0, 4π}, {y, 0, 4π}
]
```

6.6

The hard part is parameterizing the square. Use the parameterization developed for the billiard trajectories of Chap. 2.

```
ParametricPlot[
 {1/2 - Abs[Mod[t, 2] - 1],
  1/2 - Abs[Mod[1/2 + t, 2] - 1]},
 {t, 0, 2}
]
```

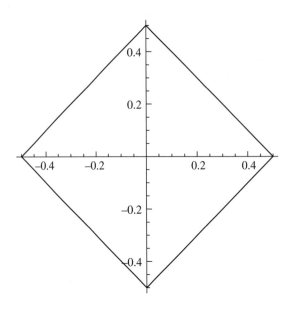

6.7

Rotate the original square by an angle of θ, then translate it to the right by r.

```
r = 1;
Manipulate[

 RevolutionPlot3D[

  {r + (1/2 - Abs[-1 + Mod[t, 2]]) Cos[θ] -
   (1/2 - Abs[-1 + Mod[1/2 + t, 2]]) Sin[θ],
   (1/2 - Abs[-1 + Mod[1/2 + t, 2]]) Cos[θ] +
   (1/2 - Abs[-1 + Mod[t, 2]]) Sin[θ]}, {t, 0, 2},

  PerformanceGoal → "Quality"],

  {θ, 0, Pi / 2}

]
```

6.8

```
RegionPlot3D[z >= 0 && z <= 5 (1 - Sqrt[x^2 + y^2]),
  {x, -1, 1}, {y, -1, 1},
  {z, 0, 5}
]
```

6.9

Let the cone point be at (a,b,c). Now draw lots of lines connecting the cone point to the circular base of the cone. Use **Manipulate** to move the cone point.

It is possible to describe the points (x,y,z) that lie in the cone with inequalities, but **RegionPlot3D** gives unsatisfactory results.

```
Manipulate[
  Graphics3D[
    Table[
      Line[{{a, b, c}, {Cos[t], Sin[t], 0}}]
      , {t, 0.0, 2 Pi, 2 Pi / 50.0}
    ],
    PlotRange → {{-2, 2}, {-2, 2}, {0, 5}}
  ],
  {a, -2, 2},
  {b, -2, 2},
  {{c, 5}, 0, 5}
]
```

6.10

```
ParametricPlot3D[
  {Cos[3 t], Cos[5 t + Pi / 4], Cos[7 t + Pi / 12]},
  {t, 0, 2 Pi},
  PlotStyle → Tube[.03]
]
```

Quiz 7

7.1

```
Limit[ Sin[x²] / x , x → 0]
```

0

7.2

```
f[x_] := -x^3 + x + 1
m = f'[1]

-2

(* plot curve and tangent line *)
Plot[{f[x], f[1] + m (x - 1)}, {x, -1, 2}]
```

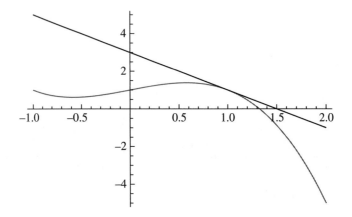

7.3

```
f[x_] := -x^3 + x + 1;
Manipulate[
 m = f'[a];
 Plot[{f[x], f[a] + m (x - a)}, {x, -2, 2},
  PlotStyle → {Black, Blue},
  PlotRange → {{-2, 2}, {-2, 2}},
  Epilog -> {
    Red, PointSize[.015], Point[{a, f[a]}]
   }
 ],
 {{a, 0}, -2, 2}
]
```

7.4

```
solution = N[
  Minimize[{(x - 3)² + (y - 5)², x² - y² == 1}, {x, y}]
]
```

```
{2.25727, {x → 4.07871, y → 3.95422}}
```

```
Plot[{√(x² - 1), -√(x² - 1)}, {x, -5, 5},
  AspectRatio → Automatic,
  Epilog → {
    PointSize[.02],
    Point[{3, 5}],
    Point[{x, y} /. solution[[2]]]
  }
]
```

7.5

```
(* {x₁,y₁} is on the circle,
{x₂,y₂} is on the hyperbola *)
solution = N[
   Minimize[{(x₁ - x₂)² + (y₁ - y₂)²,
      (x₁ - .5)² + (y₁ - 1)² == .25, x₂² - y₂² == 1},
    {x₁, y₁, x₂, y₂}]
 ]
{0.0760671, {x₁ → 0.940806,
   x₂ → 1.18396, y₁ → 0.764012, y₂ → 0.633839}}
```

7.6

```
ParametricPlot[{Cosh[t], Sinh[t]}, {t, -1, 1},
  PlotRange → {{0, 2}, {0, 2}},
  AspectRatio → Automatic,
  Epilog → {
    Circle[{.5, 1}, √.25],
    Line[{{x₁, y₁}, {x₂, y₂}} /. solution[[2]]]
    }
 ]
```

7.7

Plotting the curves reveals that they intersect near –2 and 1.

```
a = x /. FindRoot[e^x == 4 - x^2, {x, -2}]
b = x /. FindRoot[e^x == 4 - x^2, {x, 1}]
```

-1.96464

1.05801

```
area = NIntegrate[4 - x^2 - e^x, {x, a, b}]
```

6.42769

7.8

```
(* find moment about x-axis *)
mx = NIntegrate[y, {x, a, b}, {y, e^x, 4 - x^2}]
```

13.4812

```
(* find moment about y-axis *)
my = NIntegrate[x, {x, a, b}, {y, e^x, 4 - x^2}]
```

-2.65234

```
(* find center of mass *)
{X, Y} = {my / area, mx / area}
```

{-0.412643, 2.09736}

```
Plot[{e^x, 4 - x^2}, {x, a, b},
 AspectRatio → Automatic,
 Epilog → {PointSize[.02], Point[{X, Y}]}
]
```

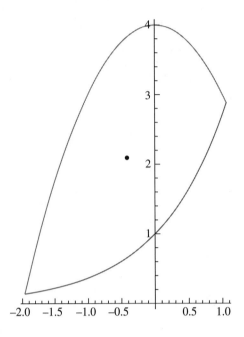

7.9

The hard part is visualizing the solid. Here is 1/16 of the solid.

```
Plot3D[√(1 - x²), {x, 0, 2}, {y, 0, 2},
 BoxRatios → {2, 2, 1},
 PlotStyle → {Opacity[.8]},
 RegionFunction →
  Function[{x, y, z}, y ≤ x && x ≤ 1],
 Filling → Bottom,
 FillingStyle -> {Green, Opacity[.6]}
]
```

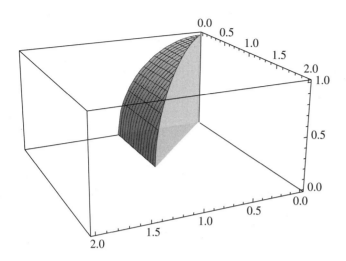

```
volume =
    16 Integrate[√(1 - x²), {y, 0, x}, {x, 0, 1}]
```

$$\frac{16}{3}$$

7.10

```
Solve[ (3 H² + 8 H R + 3 R²) / (4 H + 8 R) == H,  R]
```

$$\left\{\left\{R \to -\frac{H}{\sqrt{3}}\right\}, \left\{R \to \frac{H}{\sqrt{3}}\right\}\right\}$$

```
ArcTan[ (H/√3) / H ]
```

$$\frac{\pi}{6}$$

Quiz 8

8.1

$$\text{Solve}\left[x\,y + y - 3 == \frac{2\,x + y}{3\,x + 4},\ x\right]$$

$$\left\{\left\{x \to \frac{11 - 7\,y - \sqrt{121 - 10\,y + 13\,y^2}}{6\,y}\right\},\right.$$

$$\left.\left\{x \to \frac{11 - 7\,y + \sqrt{121 - 10\,y + 13\,y^2}}{6\,y}\right\}\right\}$$

$$\text{Solve}\left[x\,y + y - 3 == \frac{2\,x + y}{3\,x + 4},\ y\right]$$

$$\left\{\left\{y \to \frac{12 + 11\,x}{3 + 7\,x + 3\,x^2}\right\}\right\}$$

8.2

```
(* this will find roots exactly *)
Solve[x^4 + 5 x^3 - x + 6 == 0, x]

(* approximate the roots *)
NSolve[x^4 + 5 x^3 - x + 6 == 0, x]
```

$$\{\{x \to -4.90772\},\ \{x \to -1.24491\},$$
$$\{x \to 0.576316 - 0.80617\,i\},$$
$$\{x \to 0.576316 + 0.80617\,i\}\}$$

8.3

```
ListPlot[
    {Re[#], Im[#]} & /@ (x /. NSolve[x^6 + x + 1 == 0, x]),
    PlotStyle → PointSize[.02]
]
```

8.4

$$\text{Resultant}\left[x^6 + a\,x + 1, \; D\left[x^6 + a\,x + 1, \; x\right], \; x\right]$$

$$46\,656 - 3125\,a^6$$

$$\text{Solve}\left[46\,656 - 3125\,a^6 == 0, \; a\right]$$

$$\left\{\left\{a \to -\frac{6}{5^{5/6}}\right\}, \; \left\{a \to \frac{6}{5^{5/6}}\right\}, \; \left\{a \to -\frac{6\,(-1)^{1/3}}{5^{5/6}}\right\},\right.$$

$$\left.\left\{a \to \frac{6\,(-1)^{1/3}}{5^{5/6}}\right\}, \; \left\{a \to -\frac{6\,(-1)^{2/3}}{5^{5/6}}\right\}, \; \left\{a \to \frac{6\,(-1)^{2/3}}{5^{5/6}}\right\}\right\}$$

There are five roots if a is any one of the values above; six otherwise.

8.5

Since the system is small, it is easiest to use **Solve**. Here we solve for x and y in terms of z. There are infinitely many solutions.

$$\text{Solve}[\{x - z == 4, \; 2\,x + y - 3\,z == 5\}, \; \{x, \; y\}]$$

$$\{\{x \to 4 + z, \; y \to -3 + z\}\}$$

8.6

Use **NSolve** to find roots and **Select** to get the real roots.

```
realroots = Select[
  θ /. NSolve[2 Sin[3 θ] == 1 + Cos[3 θ] ^ 2, θ],
  Im[#] == 0 &
]
```

Solve::ifun : Inverse functions are being used by Solve, so some
   solutions may not be found; use Reduce for complete solution information. ≫

```
{-1.82062, -1.32097, 0.273776,
 0.773422, 2.36817, 2.86782}
```

The points where the two curves intersect are, in Cartesian coordinates,

```
points = {2 Sin[3 #] Cos[#], 2 Sin[3 #] Sin[#]} & /@
   realroots
```

```
{{-0.361973, -1.41865}, {0.361973, -1.41865},
 {1.40957, 0.395847}, {1.0476, 1.0228},
 {-1.0476, 1.0228}, {-1.40957, 0.395847}}
```

Let's plot the curves and the points.

```
PolarPlot[{2 Sin[3 θ], 1 + Cos[3 θ] ^2},
    {θ, 0, 2 π},
    Epilog → {PointSize[.03],
        Point[{#[[1]], #[[2]]}] & /@ points}
]
```

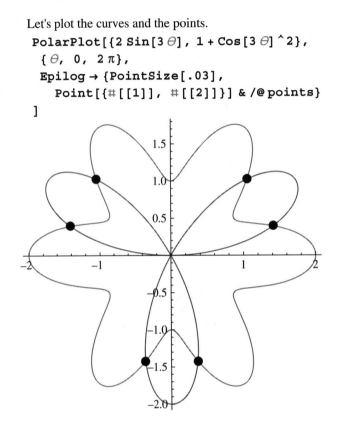

8.7

```
solution =
  DSolve[{x''[t] == -9.8, x[0] == 1000, x'[0] == 0},
    x, t]
```

$$\{\{x \to \text{Function}[\{t\}, 1000 - 4.9\, t^2]\}\}$$

Object hits ground when $x[t] = 0$.

```
Solve[x[t] == 0 /. solution, t]
```

$$\{\{t \to -14.2857\}, \{t \to 14.2857\}\}$$

8.8

```
solution =
 DSolve[{x''[t] == -0.163 x'[t] - 9.8,
   x[0] == 3000, x'[0] == 0}, x, t]
```

$$\left\{\left\{x \rightarrow Function\left[\{t\}, e^{-0.163\,t}\right.\right.\right.$$
$$\left.\left.\left.\left(-368.851 + 3368.85\,e^{0.163\,t} - 60.1227\,e^{0.163\,t}\,t\right)\right]\right\}\right\}$$

Find out when height is 2000 m.

```
Solve[x[t] == 2000 /. solution, t]
```

InverseFunction::ifun :
  Inverse functions are being used. Values may be lost for multivalued inverses. ≫

Solve::ifun: Inverse functions are being used by Solve, so some
  solutions may not be found; use Reduce for complete solution information. ≫

```
{{t → -10.3425}, {t → 22.6138}}
```

Check that height is 2000 m when $t = 22.6138$ sec.

```
x[22.6138] /. solution
{2000.}
```

8.9

Terminal velocity is limiting velocity.

```
Limit[x'[t] /. solution, t → Infinity]
{-60.1227}
```

The terminal velocity is about 60 m/sec downward.

8.10

```
solution =
 DSolve[{x'[t] == x[t] - 10 y[t],
   y'[t] == 15 x[t] + y[t]}, {x, y}, t];

Manipulate[
 ParametricPlot[
```

```
{
 {x[t], y[t]} /. solution /.
  {C[1] → pt1[[1]], C[2] → pt1[[2]]},
 {x[t], y[t]} /. solution /.
  {C[1] → pt2[[1]], C[2] → pt2[[2]]},
 {x[t], y[t]} /. solution /.
  {C[1] → pt3[[1]], C[2] → pt3[[2]]}
 },
 {t, 0, length},
 PlotRange → {{-2, 2}, {-2, 2}}
],
(* controllers *)
{{length, .25, "Length"}, 0.001, 1},
{{pt1, {RandomReal[2] - 1, RandomReal[2] - 1}},
 Locator},
{{pt2, {RandomReal[2] - 1, RandomReal[2] - 1}},
 Locator},
{{pt3, {RandomReal[2] - 1, RandomReal[2] - 1}},
 Locator}
]
```

Quiz 9

9.1

```
mostDense = Take[
  Sort[
   {#,
     CountryData[#, "Population"] /
     CountryData[#, "Area"]} & /@ CountryData[],
   #1[[2]] > #2[[2]] &
   ],
   10
  ]
```

For least dense, change 10 to −10.

9.2

```
ListPlot[
 {CountryData[#, "GDP"],
   CountryData[#, "InfantMortalityFraction"]} & /@
  CountryData[]
]
```

9.3

```
worldPop =
 Total[CountryData[#, "Population"] & /@
   CountryData[]
 ]

(* the 10 biggest countries *)
biggest = Take[
  Sort[
   {#, CountryData[#, "Population"]} & /@
    CountryData[],
   #1[[2]] > #2[[2]] &
  ],
  10
 ]

biggestPop = Total[#[[2]] & /@ biggest]

biggestPlusOther =
 Join[biggest, {{"Other", worldPop - biggestPop}}]

(* plotting the population of countries *)
Needs["PieCharts`"]
labels =
  Graphics[Text[#[[1]]], ImageSize → {80, 20}] & /@
   biggestPlusOther;
PieChart[#[[2]] & /@ biggestPlusOther,
 PieLabels → labels,
 PlotLabel →
   "Population of the Biggest Countries"
]
```

Population of the Biggest Countries

9.4

```
(* using Tooltip to bring up name of each
country *)
Graphics[
 {LightGray,
  EdgeForm[Black],
  Tooltip[
     CountryData[#, "Polygon"],
     #
    ] & /@
   CountryData["Africa"]
 }
]
```

9.5

```
Length[DictionaryLookup[]]
```

92 518

9.6

```
DictionaryLookup["b" ~~ ___ ~~ "w"]
```

9.7

Yes!
ates, east, eats, etas, sate, seat, teas

9.8

lead, load, goad, gold

9.9

Happily, there is no path from good to evil! There are no words in the
Mathematica dictionary which are only one letter change away from evil.
If we allow the word "eval," then a Word-Link exists.

9.10

```
(* define Collatz function *)
g[n_] := If[Mod[n, 2] == 0, n/2, 3 n + 1]
(* initialize vertices and edges of graph *)
vertices = {1};
edges = {1 → 4};
(* consider each integer n from 2 up to
 some value *)
Do[
 m = n;
 (* follow the course of m until we reach
  a vertex already in the graph *)
 (* append m and the edge from m to g[m]
  to the graph as we go *)
 While[Intersection[{m}, vertices] == {},
  AppendTo[vertices, m];
  AppendTo[edges, m → g[m]];
  m = g[m]
 ];
 , {n, 2, 8}
]
(* plot the graph *)
GraphPlot[edges, VertexLabeling → True,
 DirectedEdges → True]
```

Quiz 10

10.1 (a)

```
f[x_] := -x
Abs[f'[0]]
```

1

10.1 (b)

```
f[x_] := Tan[x]
Abs[f'[0]]
```

1

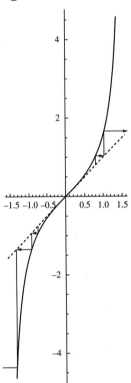

10.1 (c)

```
f[x_] := Sin[x]
Abs[f'[0]]

1
```

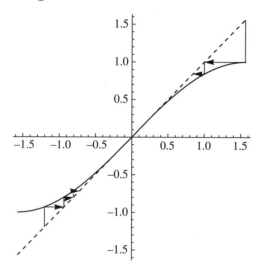

10.2

The bifurcation takes place between −1.249 and −1.251.

```
Do[
  f[x_] := x^2 + c;
  tailOfOrbit = NestList[f, Nest[f, 0, 10 000],
    8];
  Print["c=", c, " ", tailOfOrbit],
  {c, -1.249, -1.251, -.002}
]
c=-1.249 {0.206399, -1.2064, 0.206399, -1.2064,
  0.206399, -1.2064, 0.206399, -1.2064, 0.206399}
c=-1.251 {0.180917, -1.21827, 0.23318, -1.19663,
  0.180917, -1.21827, 0.23318, -1.19663, 0.180917}
```

10.3

```
f[x_] := x^2 + c
(* find fixed points of f[f[x]] which are
 not fixed points of f[x] *)
fix1 = x /. Solve[f[x] == x, x];
fix2 = x /. Solve[f[f[x]] == x, x];
fix = Complement[fix2, fix1]
```

$$\left\{ \frac{1}{2} \left(-1 - \sqrt{-3 - 4\,c} \right), \frac{1}{2} \left(-1 + \sqrt{-3 - 4\,c} \right) \right\}$$

```
(* find slope of f[f[x]] at the period-
 2 points *)
Simplify[D[f[f[x]], x] /. x → # & /@ fix]
```

$\{4\,(1+c),\ 4\,(1+c)\}$

```
(* determine value of c when the slope is -1 *)
Solve[4 (1 + c) == -1, c]
```

$$\left\{ \left\{ c \rightarrow -\frac{5}{4} \right\} \right\}$$

10.4

The bifurcation takes place between $c = -1.367$ and $c = -1.369$.

```
Do[
  f[x_] := x^2 + c;
  tailOfOrbit = NestList[f, Nest[f, 0, 10 000],
    8];
  Print["c=", c, " ", tailOfOrbit],
  {c, -1.367, -1.369, -.002}
 ]
c=-1.367 {-0.0822364, -1.36024,
   0.483245, -1.13347, -0.0822364,
   -1.36024, 0.483245, -1.13347, -0.0822364}
c=-1.369 {-0.0623327, -1.36511,
   0.494538, -1.12443, -0.104652,
   -1.35805, 0.475294, -1.1431, -0.0623327}
```

Answers

10.5

In all the following $c = -0.1 + 0.8$ i and xdiv = ydiv = 1000. In the two plots, the max number of iterations in orbitLength is 25 and 500, respectively.

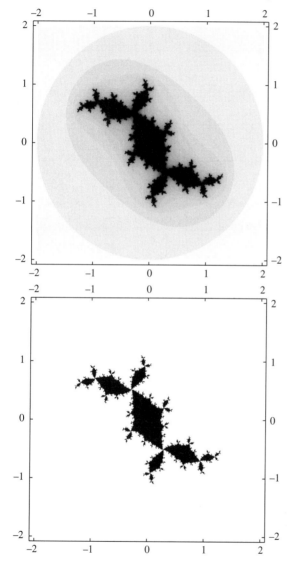

10.6

This will make the frames of the movie.

```
frames = Table[
  ArrayPlot[JuliaData[ .8 E^(I t),
    {{-2, 2}, {-2, 2}}, 1000, 1000],
   DataRange → {{-2, 2}, {-2, 2}},
   FrameTicks → Automatic
  ],
  {t, 0,  2 Pi,  Pi / 10}
 ]
```

10.7

Zooming in on the point {x, y} in the Mandelbrot Set. Each frame of the movie is .9 times the size of the previous frame.

```
{x, y} = {-.1515, 1.025};
size[k_] := .9^k;
Do[
 Print[
  MandelbrotPlot[
   MandelbrotData[{{x - size[k], x + size[k]},
     {y - size[k], y + size[k]}}, 300], grayScheme,
    5, 5
   ]
  ],
  {k, 0, 20}
 ]
```

10.8

Changing the function to `f[z] = z⁴ + c` gives the following "Mandelbrot" Set.

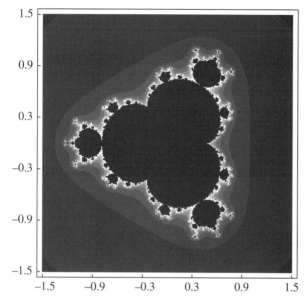

Quiz 11

11.6

```
CreatePalette[
 PasteButton[
   ∞
  ]
 ]
```

11.7

```
CreatePalette[
 PasteButton[
  DisplayForm[
   RowBox[
    {"(",
     GridBox[
      {{■, □, □}, {□, □, □}, {□, □, □}}
     ],
     ")"
    }
   ]
  ]
 ]
 ,
 WindowTitle → "Matrix"
]
```

Final Exam

1. Parenthesis, (and), are used for grouping algebraic expressions, such as $3(x + 1)$. Square brackets, [and], are used to surround the arguments of functions, such as **Range[5]**. Double square brackets, [[and]], are used to access the n-th element of a list as in **list[[3]]**. Braces, { and }, are used to delimit lists.

2. A semicolon placed after a command will suppress the output from that command. Double clicking an output cell bracket will hide the cell.

3. The expression $\sqrt{3}$ represents the square root of three exactly. Using $\sqrt{3.0}$ will give a decimal approximation.

4. The numerical function **N[*expr*]** will give a numerical approximation of *expr*.

5. Evaluating **?***FunctionName* will bring up information about that function with a hyperlink to the Help Files.

6. Evaluating **?*String** will bring up information on all functions whose names end in **String**.

7. Evaluating *expr /. lhs → rhs* will replace every occurrence of *lhs* in *expr* with *rhs*.

8. The construction *lhs = rhs* will immediately replace *lhs* with the current value of *rhs*. The value of *lhs* is now fixed at this value. Using *lhs := rhs* instead will cause *lhs*, whenever it is used, now or in the future, to be replaced at the time that it is used by the value of *rhs* at that time.

9. The ampersand, &, is used in the construction of pure functions. For example $\sharp^2$ &, defines a pure function that will square its input.

10. The at symbol, @, can be used in at least two different ways. First, we may replace *Function[expr]* with *Function@expr* as in **MatrixForm@Table[i+j, {i, 1, 3}, {j, 1, 3}]** instead of **MatrixForm[Table[i+j, {i, 1, 3}, {j, 1, 3}]]**. Second, **Map[*function, expr*]** can be replaced by *function/@expr*.

11. The square root function (like most *Mathematica* functions) is *listable*. The square root of a list of elements is the list of the square roots of the elements.

12. The first construction, {1, 2, 3}*{*a, b, c*} will give a list of the products of the respective elements from each list, namely, {*a*, 2*b*, 3*c*}. This is because multiplication is listable. In the second construction the dot stands for matrix multiplication, so this gives

$$\begin{pmatrix} 1 & 2 & 3 \end{pmatrix} \begin{pmatrix} a \\ b \\ c \end{pmatrix} = a + 2b + 3c.$$

13. & 14.

```
Plot[
  {x², 10 - x²}, {x, -3, 3},
  Filling → {1 → {{2}, {Yellow, None}}}
]
```

15.

```
f[x_] := x^3 - 5 x + 1
Manipulate[
 (* find points on curve and eliminate
  complex points *)
 solutions = Select[
   x /. NSolve[
     f[x] - pt[[2]] == f'[x] (x - pt[[1]]), x
    ],
   Im[#] == 0 &
  ];
 (* plot curve and use Epilog to put in lines *)
 Plot[f[x], {x, -4, 4},
  Epilog -> {Line[{pt, {#, f[#]}}] & /@ solutions}
 ],
 {{pt, {3, 1}}, Locator}
]
```

16.

```
ParametricPlot[{y^2, y}, {y, -4, 4}]
```

17.

```
Plot3D[Abs[x] Abs[y], {x, -1, 1}, {y, -1, 1},
 RegionFunction -> Function[{x, y, z}, x^2 + y^2 < 1]
]
```

18.

```
RevolutionPlot3D[Sqrt[x - 1], {x, 0, 3}]
```

19.

```
ContourPlot[Sin[x^2 - y^3], {x, -1, 1}, {y, -1, 1}]
```

20.

```
f[x_, y_, z_] :=
 (z - Sqrt[1 + x^2 + y^2]) (x^2 + y^2 + z^2 - 9)
ContourPlot3D[
 f[x, y, z], {x, -5, 5}, {y, -5, 5}, {z, -5, 5},
 Contours → {3}
]
```

21.

```
In[58]:= (* four nice points that are equidistant
    apart are {1,0,0},{0,1,0},{0,0,1},
    {1,1,1} *)
    vertices = a {{1, 0, 0}, {0, 1, 0}, {0, 0, 1},
       {1, 1, 1}};
    (* find a so that the distance between two
    vertices is 10 *)
    Solve[Norm[vertices[[1]] - vertices[[2]]] == 10,
     a]
```

$$Out[59]= \left\{\left\{a \to -5 \sqrt{2}\right\}, \left\{a \to 5 \sqrt{2}\right\}\right\}$$

```
In[348]:= (* set the value of a *)
    a = 5 √2 ;
```

```
In[349]:= (* create list of edges *)
    (* use Flatten to remove grouping caused
    by Table *)
    edges = Flatten[
       Table[{vertices[[i]], vertices[[j]]},
        {i, 1, 3}, {j, i + 1, 4}],
       1
      ];
```

```
In[350]:= (* plot the tetrahedron *)
    Graphics3D[
     {Sphere[#, 1] & /@ vertices,
      Cylinder[#, .25] & /@ edges}
     ]
```

22.

$$\text{vol} = 1 / 3 \ \pi \ r^2 \ h$$

$$\text{area} = \pi \ r^2 + \pi \ r \ \sqrt{r^2 + h^2}$$

(* arbitrarily set area to π to find shape
 of optimal cone *)

$$\text{Maximize}[\{\text{vol}, \ \text{area} == \pi\}, \ \{r, \ h\}]$$

Out[97]= $\dfrac{1}{3} \ h \ \pi \ r^2$

Out[98]= $\pi \ r^2 + \pi \ r \ \sqrt{h^2 + r^2}$

Out[99]= $\left\{ \dfrac{\pi}{6 \sqrt{2}}, \ \left\{ r \to \dfrac{1}{2}, \ h \to \sqrt{2} \right\} \right\}$

In[100]:= (* ratio of height to radius for optimal
      cone is *)

$$\sqrt{2} \ \Big/ \ (1 / 2)$$

Out[100]= $2 \sqrt{2}$

23.

In[120]:= (* assume density δ is 1 unit *)
      (* dimensions are in meters *)
      (* center of mass lies on vertical line
       below cone point due to symmetry *)

$\delta = 1;$

$s = 230;$

$h = 147;$

$\text{mass} = 1 / 3 \ s^2 \ h \ \delta$

Out[123]= 2 592 100

In[124]:= (* similar triangles gives that side length
      of horizontal slice at height z is 115/147
      (147-z) *)
      (* find moment wrt to xy-plane by
        integrating mass of horizontal slice at
        height z times moment arm of z *)

```
m_xy = Integrate[(115 (147 - z) / 147)^2 δ z,
    {z, 0, 147}]
```

Out[124]= $\dfrac{95\,259\,675}{4}$

In[125]:= (* find height of center of mass *)
```
zbar = N[m_xy / mass]
```

Out[125]= 9.1875

24.

In[126]:= (* Let T[t] be temperature of pizza at time
    t *)
(* solve differential equation with unknown
    decay constant *)
```
sol = DSolve[
    {T'[t] == k (T[t] - 72), T[0] == 450}, T, t
    ]
```

Out[126]= $\left\{\left\{T \to \text{Function}\left[\{t\}, 18\left(4 + 21\,e^{k\,t}\right)\right]\right\}\right\}$

In[127]:= (* solve for constant *)
```
constant = Solve[T[5] == 300 /. sol, k][[1]]
```

Solve::ifun : Inverse functions are being used by Solve, so some
    solutions may not be found; use Reduce for complete solution information. »

Out[127]= $\left\{k \to -\dfrac{1}{5}\,\text{Log}\left[\dfrac{63}{38}\right]\right\}$

In[128]:= (* knowing constant,
    find time when Pizza is 200 degrees *)
```
NSolve[
    (T[t] == 200 /. sol /. constant)[[1]], t
    ]
```

Solve::ifun : Inverse functions are being used by Solve, so some
    solutions may not be found; use Reduce for complete solution information. »

Out[128]= $\{\{t \to 10.7098\}\}$

25.

```
primeList = Table[Prime[i], {i, 1, 1000}];
Select[primeList, Mod[#, 4] == 1 &];
```

26.
```
twins = {};
i = 1;
While[
 Length[twins] < 100,
 {x, y} = {Prime[i], Prime[i + 1]};
 If[
  y - x == 2, AppendTo[twins, {x, y}]
 ];
 i++
]
twins
```

27.
```
In[144]:= (* decides if n is a perfect number *)
      perfectQ[n_] := Module[
        {sum = 0},
        (* find sum of proper divisors *)
        Do[If[Mod[n, k] == 0, sum += k]
         , {k, 1, n / 2}];
        sum == n
       ]

In[153]:= (* find first four perfect numbers *)
      data = {};
      n = 1;
      While[Length[data] < 4,
       If[perfectQ[n], AppendTo[data, n]];
       n++
      ]
      data

Out[156]= {6, 28, 496, 8128}
```

28.
```
Print[a = 0]
Print[b = 1]
Do[
 Print[c = a + b];
 a = b;
 b = c,
 {i, 1, 100}
]
```

29.
```
k = 1;
While[2^k < 10^6, Print[2^k]; k++]
```

30.
```
(* h[t] is height of ball at time t *)
h[t_] := -16 t^2 + 100;
r = 2; (* radius of ball *)
(* produce frames of movie *)
frames = Table[
   Graphics[
     {{Thickness[.05], Line[{{-10, 0}, {10, 0}}]},
      Disk[{0, h[t] + r}, r]},
     PlotRange → {{-10, 10}, {-1, 110}}
     ],
     {t, 0, 5/2, .05}
   ];
(* use ListAnimate to view movie *)
 ListAnimate[
    frames
 ]
```

31.
```
(* use Select to cull out the palindromes *)
Select[
 DictionaryLookup[],
 StringJoin@Reverse@Characters[#] == # &
 ]
```

32.
```
In[250]:= (* total area of all countries in square km *)
         land =
          Total[CountryData[#, "Area"] & /@ CountryData[]]
```
$$Out[250]= 1.35885 \times 10^8$$

```
In[264]:= (* radius of earth in km *)
         r = 6370;
         (* surface area *)
         totalArea = N[4 π r^2]
```
$$Out[265]= 5.09904 \times 10^8$$

```
In[263]:= (* fraction of surface covered by water *)
         1 - land / totalArea
```
$$Out[263]= 0.733508$$

33.
```
Graphics[
 {LightBlue,
  EdgeForm[Black],
  Tooltip[
     CountryData[#, "Polygon"],
     #
    ] & /@
   CountryData["Africa"]
 }
]
```

34.
```
gdpData = {#[[1, 1]], #[[2]] / 10^12} & /@
   CountryData["UnitedStates",
    {"GDP", {1988, 2008}}];
gdpPlot = ListPlot[gdpData,
   PlotRange → {{1988, 2008}, {5, 13}},
   AxesOrigin → {1988, 5},
   PlotLabel →
    "GDP of USA in trillions of dollars."
   ]
```

35.

In[330]:= (* fit parabola to data *)
      poly = Fit[gdpData, {1, x, x^2}, x]
      fitPlot = Plot[poly, {x, 1988, 2008}];
      Show[gdpPlot, fitPlot]

Out[330]= $41\,722.1 - 42.2105\,x + 0.0106771\,x^2$

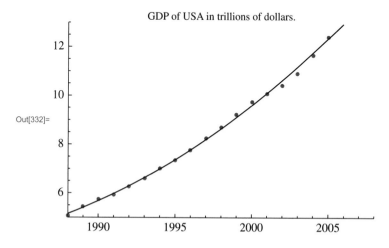

GDP of USA in trillions of dollars.

In[333]:= (* projected value of GDP in 2013 in
      trillions of dollars *)
      poly /. x → 2013

Out[333]= 17.8715

INDEX

Wolfram *Mathematica* ®6
PRODUCT TRIAL

This book includes a 30-day *Mathematica*® product trial. To download your trial, go to http://www.wolfram.com/books/resources and enter the license number below (/optional:/ to be guided through the installation process).

Trial license number: L3251-0001